Wine Fermentation

Wine Fermentation

Special Issue Editor

Harald Claus

MDPI • Basel • Beijing • Wuhan • Barcelona • Belgrade

MDPI

Special Issue Editor
Harald Claus
Johannes Gutenberg-University of Mainz
Germany

Editorial Office
MDPI
St. Alban-Anlage 66
4052 Basel, Switzerland

This is a reprint of articles from the Special Issue published online in the open access journal *Fermentation* (ISSN 2311-5637) from 2018 to 2019 (available at: https://www.mdpi.com/journal/fermentation/special_issues/wine_fermentation)

For citation purposes, cite each article independently as indicated on the article page online and as indicated below:

LastName, A.A.; LastName, B.B.; LastName, C.C. Article Title. *Journal Name* **Year**, *Article Number, Page Range.*

ISBN 978-3-03897-674-5 (Pbk)
ISBN 978-3-03897-675-2 (PDF)

Cover image courtesy of Harald Claus.

Contents

About the Special Issue Editor

Harald Claus, Dr. rer. nat., born in Mainz (Germany), studied biology at the Technical University of Darmstadt and completed his dissertation at the Institute of Microbiology. He received his doctorate in 1983 and worked as a scientist in the private sector in the following years. In 1986, he received an industrial scholarship and worked as a microbiologist at the Institute for Water, Soil, and Air Hygiene of the German Federal Health Office. Since 1998, he has been group leader and lecturer at the Institute of Molecular Physiology–Microbiology and Wine Research at the Johannes Gutenberg University in Mainz. In the course of his scientific career, he has gained experience in various research topics (https://www.researchgate.net/profile/Harald_Claus/publications). Current interests include: Biotechnological enzymes; phenoloxidases; yeasts; bioremediation; bio-control; wine microbiology.

fermentation

MDPI

Editorial

Wine Fermentation

Harald Claus

Institute of Molecular Physiology, Microbiology and Wine Research, Johannes Gutenberg-University of Mainz, Becherweg 15, D-55099 Mainz, Germany; hclaus@uni-mainz.de; Tel.: +49-6131-3923542

Received: 2 February 2019; Accepted: 9 February 2019; Published: 12 February 2019

Keywords: extraction methods; color intensity; phenolic content; *Saccharomyces*; *Lachancea*; yeast hybrids; metabolomics; sulfur compounds; oenological enzymes; process control

Currently wineries are facing new challenges due to actual market demands for creation of products exhibiting more individual flavors. Serious climate changes have provoked a search for grape varieties with specific features such as convenient maturation times, enhanced tolerances towards dryness and osmotic stress as well as resistance against invasive plant-pathogenic organisms. The next generation of yeast starter cultures should produce wines with an appealing sensory profile and less alcohol. This Special Issue comprises actual studies addressing some problems and solutions for environmental, technical and consumer challenges of winemaking today. The contributions are focused on modern techniques and approaches at different stages of fermentation.

The development of new sophisticated mass-spectroscopic methods has enabled considerable progress in chemical analysis in recent years. It allows the identification of the major part of chemical structures, at best the entire metabolite spectrum of an organism. Pinu [1] gives an overview of how metabolome analyses enable the determination of the geographical origin of a grape or tracking of yeast-specific characteristics. Analysis can also be limited to distinct substance classes and assists elucidation of corresponding biosynthetic pathways.

One of these specific chemicals in wine are reduced sulfur compounds which usually produce unpleasant off-odors and make such wines often unsaleable. Müller and Rauhut [2], outstanding experts in this field, report on the origin and nature of such substances and the complex chemical reactions that they undergo during wine storage.

Apart from gustatory pleasures and occasional stimulating effects, moderate wine consumption has been recognized as beneficial to human health in many clinical studies. In particular, polyphenols in red wine are associated with positive antioxidant and cardiovascular properties. Color intensity is the first decisive quality feature for the consumer. Modern winemaking techniques take care to maintain high levels of these desirable compounds. Sommer and Cohen [3] applied six different physico-chemical treatments (e.g., ultrasound and microwave-assisted extraction) for effective and sustainable color extraction from eleven red grape varieties. They concluded that color characteristics of the finished product cannot easily be predicted from the initial extraction success but depend on the specific anthocyanin spectrum of the individual cultivar used.

Moreover, the maturity status of the fruit can exert a significant influence on extraction efficiency and color stability. Casassa et al. [4] showed that microwave treatment leads to increased phenol and long-term color levels particularly in wines produced from unripe grapes and less in those derived from ripe grapes.

Claus and Mojsov [5] summarize knowledge about current applications of technical enzymes in winemaking. They stress that, although mostly obtained from mushrooms, many wine-associated microorganisms produce enzymes of oenological interest. These biocatalysts could be used either as enzyme formulations or directly in the form of starter cultures to increase juice yields, color intensity and aroma of wine.

Saccharomyces cerevisiae is and certainly will remain the primary yeast for wine fermentations. Nevertheless, König and Claus [6] report that non-conventional *Saccharomyces* species like *Saccharomyces bayanus*, *Saccharomyces kudriavzevii* and their natural hybrids are of increasing interest as they exhibit good fermentative capacities, producing wines with lower ethanol and higher glycerol concentrations. In this way, the increased sugar content of the grapes due to global warming could be counteracted. In addition, they may be tools to avoid stuck fermentations under nitrogen limitations. Accordingly, Kelly et al. [7] demonstrated the potential of adapted autochthonous yeasts such as strains of *S. bayanus* to produce individual wines even in cool climate regions.

Non-*Saccharomyces* yeasts, considered essentially "wild" spoiling microorganisms in the past, are seen as beneficial today as they can improve the wine sensory profile, especially when grown in controlled mixed starter fermentations together with *S. cerevisiae*. Vilela [8] reviews data about *Lachancea thermotolerans* for wine production, which is characterized by reduced levels of alcohol and volatile acids in favor of high concentrations of the flavor compound ethyl lactate. Thus, natural and artificial *Saccharomyces* hybrids as well as collections of adapted wild isolates from various ecological niches all over the world will further extend a winemaker's toolkit, allowing specific fermentations.

Wine quality can be improved by post-harvest physico-chemical or biological measures, as mentioned above, but of course also at the pre-harvest stage by appropriate winegrowing techniques. One of these methods, the so-called *cluster thinning*, was evaluated by Mawdsley et al. [9] exemplarily with a *Pino Noir* grape. Surprisingly, the authors found no quality increase in the phenolic profile irrespective of the vegetation period (cold or warm).

An important factor of consistently high product quality is process control of wine fermentations. Temperature plays a decisive role, which can vary significantly in different areas of particularly large jacketed fermentation tanks and therefore is difficult to measure and to control. Schmidt et al. [10] present an innovative open-source software program designed for the solution of this basic problem.

Acknowledgments: The editor thanks all authors and the editorial staff who contributed to the success of this Special Issue. On the occasion of his retirement, H.C. would also like to thank all members of the Institute for Microbiology and Wine Research. Special thanks go to Martina Schlander for her years of invaluable technical assistance and to Helmut König, who made him curious about the world of (wine) microbes.

Conflicts of Interest: The author declares no conflicts of interest.

References

1. Pinu, F.R. Grape and Wine Metabolomics to Develop New Insights Using Untargeted and Targeted Approaches. *Fermentation* **2018**, *4*, 92. [CrossRef]
2. Müller, N.; Rauhut, D. Recent Developments on the Origin and Nature of Reductive Sulfurous Off-Odours in Wine. *Fermentation* **2018**, *4*, 62. [CrossRef]
3. Sommer, S.; Cohen, S.D. Comparison of Different Extraction Methods to Predict Anthocyanin Concentration and Color Characteristics of Red Wines. *Fermentation* **2018**, *4*, 39. [CrossRef]
4. Casassa, L.F.; Sari, S.E.; Bolcato, E.A.; Fanzone, M.L. Microwave-Assisted Extraction Applied to *Merlot* Grapes with Contrasting Maturity Levels: Effects on Phenolic Chemistry and Wine Color. *Fermentation* **2019**, *5*, 15. [CrossRef]
5. Claus, H.; Mojsov, K. Enzymes for Wine Fermentation: Current and Perspective Applications. *Fermentation* **2018**, *4*, 52. [CrossRef]
6. König, H.; Claus, H. A Future Place for *Saccharomyces* Mixtures and Hybrids in Wine Making. *Fermentation* **2018**, *4*, 67. [CrossRef]
7. Kelly, J.; Yang, F.; Dowling, L.; Nurgel, C.; Beh, A.; Di Profio, F.; Pickering, G.; Inglis, D.L. Characterization of *Saccharomyces bayanus* CN1 for Fermenting Partially Dehydrated Grapes Grown in Cool Climate Winemaking Regions. *Fermentation* **2018**, *4*, 77. [CrossRef]
8. Vilela, A. *Lachancea thermotolerans*, the Non-*Saccharomyces* Yeast that Reduces the Volatile Acidity of Wines. *Fermentation* **2018**, *4*, 56. [CrossRef]

9. Mawdsley, P.F.W.; Dodson Peterson, J.C.; Casassa, L.F. F. Agronomical and Chemical Effects of the Timing of Cluster Thinning on *Pinot Noir* (Clone 115) Grapes and Wines. *Fermentation* **2018**, *4*, 60. [CrossRef]

10. Schmidt, D.; Freund, M.; Velten, K. End-User Software for Efficient Sensor Placement in Jacketed Wine Tanks. *Fermentation* **2018**, *4*, 42. [CrossRef]

fermentation

MDPI

Review

Grape and Wine Metabolomics to Develop New Insights Using Untargeted and Targeted Approaches

Farhana R Pinu

The New Zealand Institute for Plant and Food Research Limited, Private Bag 92169, Auckland 1142, New Zealand; farhana.pinu@plantandfood.co.nz; Tel.: +64-9926-3565

Received: 28 October 2018; Accepted: 5 November 2018; Published: 7 November 2018

Abstract: Chemical analysis of grape juice and wine has been performed for over 50 years in a targeted manner to determine a limited number of compounds using Gas Chromatography, Mass-Spectrometry (GC-MS) and High Pressure Liquid Chromatography (HPLC). Therefore, it only allowed the determination of metabolites that are present in high concentration, including major sugars, amino acids and some important carboxylic acids. Thus, the roles of many significant but less concentrated metabolites during wine making process are still not known. This is where metabolomics shows its enormous potential, mainly because of its capability in analyzing over 1000 metabolites in a single run due to the recent advancements of high resolution and sensitive analytical instruments. Metabolomics has predominantly been adopted by many wine scientists as a hypothesis-generating tool in an unbiased and non-targeted way to address various issues, including characterization of geographical origin (*terroir*) and wine yeast metabolic traits, determination of biomarkers for aroma compounds, and the monitoring of growth developments of grape vines and grapes. The aim of this review is to explore the published literature that made use of both targeted and untargeted metabolomics to study grapes and wines and also the fermentation process. In addition, insights are also provided into many other possible avenues where metabolomics shows tremendous potential as a question-driven approach in grape and wine research.

Keywords: winemaking; metabolite profiling; non-targeted analysis; classical chemical analysis; metabolic modelling; yeast physiology and metabolism; vineyard management

---

1. Introduction

Targeted metabolite analysis of grape juice and wine has been carried out for a long time, specifically after the development of gas chromatography and mass spectrometry (GC-MS) [1]. Most of these studies were performed to determine the variety of wine based on its aroma composition [2–8]. Some studies have also focused on the overall composition of grape juice and determined mainly the amount of sugars, amino acids and some important carboxylic acids using different enzymatic methods or high pressure liquid chromatography (HPLC) [9–12]. By using classical chemical analytical methods, it was only possible to determine the specific groups of metabolites that were present usually in high concentration in both grape juices and wines. The unavailability of appropriate analytical instruments and suitable methods to determine the concentrations of lower abundant metabolites were the main reasons behind this scenario [13] Therefore, the exact contribution of many significant but low-concentration metabolites to the wine fermentation process was not identified. On the contrary, comprehensive and unbiased approaches of metabolomics are now providing thorough information about many different groups of compounds in grape juices and wines and are therefore more advantageous than traditional targeted analysis [13,14].

As one of the most newly introduced "-omic" technologies, metabolomics was initially proposed as a tool in functional genomics [14]. The other "omics", technologies: genomics, transcriptomics and proteomics are focused on genes, RNA and proteins, respectively, whereas metabolomics is the study of the most downstream products of cells called metabolites [13] (Figure 1). Metabolomics is typically known as an unbiased, non-targeted and holistic analysis of cell metabolites [15]. However, application of targeted metabolomics analysis is also on the rise [16–19]. It is now an emerging research area with application in different fields including functional genomics and systems biology [14,18,20–26].

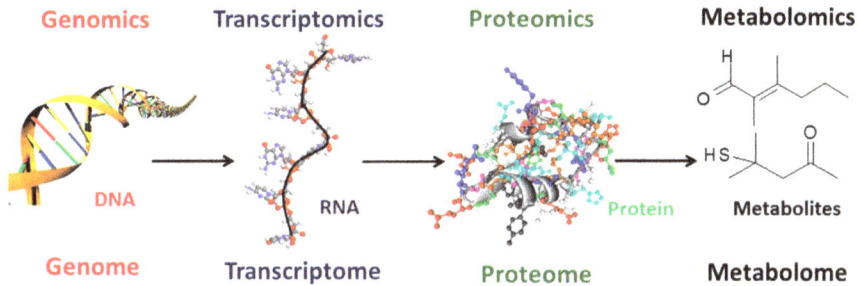

Genomics **Transcriptomics** **Proteomics** **Metabolomics**

DNA RNA Protein Metabolites

Genome **Transcriptome** **Proteome** **Metabolome**

Figure 1. The hierarchy of "omics" technologies. The post-genomics approaches (e.g., transcriptomics, proteomics, and metabolomics) together can provide comprehensive information and better understanding about the biological system (adopted from Pinu [13]).

As a question and data-driven approach, metabolomics already shows tremendous potential in food and agricultural sciences although the application of metabolomics has started just over a decade ago [27,28]. Within this time period, it has proven to be an important and powerful approach and has been used to analyze metabolites in agricultural (and food) products in both targeted and untargeted ways [16,29–33]. Like other food and agricultural products, the introduction of metabolomics in grape and wine research also garnered considerable attraction, mostly as a hypothesis-generating tool [30,34–39]. The main aim of this review is to re-visit the available published literature where either targeted or untargeted metabolomics has been applied to study grapes, wines and microorganisms associated with winemaking. Existing challenges and ways to overcome those are also provided in addition to discussing the future perspectives of metabolomics in grape and wine research.

2. Advancements in Metabolomics as an Emerging Tool within the Last Decade

The main difference among genomes, transcriptomes, proteomes and metabolomes is their chemical diversity (Figure 1) [13]. Both transcriptomes and genomes provide information on the polymeric molecules composed of only four bases, while proteomes deal with the analysis of proteins that are developed by 20 different amino acids. In contrast, the metabolome is exceptionally chemically diverse and contains 1000 to 200,000 different chemical structures [40,41]. Moreover, metabolites are the downstream products of cell metabolism and provide links with many diverse pathways that happen within a cell [13]. Many metabolites are often produced at the same time and the same metabolite can have roles in multiple pathways [42]. Metabolites produced by the cells often provide phenotypic information of the cells in response to different environmental and genetic changes [43]. Therefore, metabolite analysis is very important and provides an integrative overview of the cellular metabolism and phenotypic characteristics of the cells [13].

The metabolomics community has adopted two different ways of determining metabolome of any biological sample. Metabolite profiling is one of them and it is one of the most powerful approaches that is mainly used for untargeted metabolite analysis. In general, an untargeted metabolite profile usually contains information about both identified and unknown compounds [44]. Recently, targeted analyses of metabolites have become popular and are often combined with untargeted metabolomics data. A comparison between untargeted metabolite profiling and targeted analysis is given in Table 1. Both of these tactics are extensively used for the metabolite analysis of complex samples such cells, blood, urine and beverages.

Table 1. Different approaches of metabolite analysis.

Approach	Advantages	Disadvantages
Targeted metabolite analysis	• Low limit of detection • Usually quantitative • Data analysis and interpretation are easier • Metabolite data can be connected with pathways	• Limited number of compounds can be targeted • Non-targeted compounds are not considered • Purified standards of targeted compounds are required for quantification
Untargeted metabolite profiling	• Unbiased and comprehensive • High-throughput • Allows the discovery of new compounds not expected to be in the sample or not expected to be associated with the biological question	• Semi-quantitative • Larger number of false positives and false negatives • Many unknowns • Data interpretation can be challenging

This information was collated from [13,15,16,18].

2.1. Development of Sensitive and Reproducible Separation and Detection Techniques

During the last 10 years, the field of metabolomics has achieved a very significant improvement in terms of the analytical capability, particularly MS technologies. Now, it is possible to measure as many metabolites as possible using only a minimal amount of samples with high-throughput and exceptional sensitivity [45]. In mass spectrometry, samples can be introduced in different ways and sometimes chromatographic separation (e.g., GC, liquid chromatography and capillary electrophoresis) is a preferred method to allow maximum separation of metabolites in a complex biological sample [13]. However, direct infusion (DI) is also widely used for metabolite profiling, which is usually referred to as metabolic footprinting or fingerprinting depending on whether the analysis is of extra- or intracellular metabolites [43,46]. Due to the development of interfacing systems like atmospheric pressure ionization (API), DI-MS can be used to analyze a sample to obtain mass spectra of metabolites within a few seconds [43]. The requirement for a small amount of sample is the major advantage of using DI-MS. Moreover, no derivatization is required for this analysis and more metabolites are detected by DI-MS compared to GC-MS, making this technique best suited for high throughput non-targeted metabolite profiling [47]. However, DI-MS shows poor reproducibility when analyzing complex mixtures due to the matrix effect. The identification of metabolites by DI-MS is also very troublesome and stereoisomers cannot be resolved using this technique [43,48,49].

A variety of ion sources are available for MS: electrospray ionization (ESI), electron impact ionization (EI), chemical impact ionization (EI), matrix assisted desorption ionization (MALDI), thermospray ionization, atmospheric pressure chemical ionization (APCI), fast atom bombardment (FAB) ionization, field desorption ionization, etc. Among these, EI and ESI are the most commonly used in metabolomics [50]. The mass analyzers that have also advanced significantly and that are widely used by the scientific community are: quadrupole (Q), quadrupole ion-trap (QIT), time of flight (ToF), orbitrap, ion mobility spectrometry (IMS) and fourier transform ion cyclotron resonance (FT-ICR). Quadrupole mass analyzers are very robust, low cost and simple to use, but they offer lower mass resolution and accuracy compared to other mass analyzers [43]. On the other hand, ToF, FTICR and orbitrap, are considered extraordinary instruments that offer the highest mass resolution among all other mass analyzers [13].

NMR is another analytical instrument that has been extensively used by the metabolomics community particularly for untargeted metabolite profiling of complex mixtures (i.e., fruit juices, wines, spirits, urine and blood) [35,51–54]. The efficacy of NMR spectroscopy has been increasingly renowned for its non-invasiveness (non-destructive), throughput and linearity [55]. Moreover, NMR spectroscopy also provides structural, chemical-kinetics and other information in multidimensional applications [56]. Thus, high resolution NMR spectroscopy along with multivariate data analysis has been used for direct characterization of fruit juices, wine [54,57,58], grape berry [59,60] olive oil [61,62] and beer [63,64]. To obtain a global metabolite profile of a complex samples, NMR needs to be coupled with another non-targeted analytical approach (e.g., MS) [65]. Recently, Bruker developed and launched an instrumentation platform, scimaX MRMS, that combines the capability of NMR and MS and provides superior resolution and mass accuracy albeit the high expense.

2.2. Advancements in Data Analysis Pipelines

Mirroring the advancements of the analytical instrumentation platforms, metabolomics data analysis pipelines have also been improved significantly within last 10 years. Particularly, many efforts have been made to make data analysis more efficient and user friendly by a few prominent research groups [45,66,67]. In metabolomics, a few steps are usually involved in the whole data analysis process after raw data are generated using a suitable instrument. As such, raw data need to be preprocessed and annotated prior to statistical analysis. Post-processing steps including data filtering, imputation, normalization, data centering, scaling and transformation are also undertaken [68]. Either in-house or publicly available or commercial software or tools are used for all these steps. Therefore, data analysis generally requires a considerable amount of resources including the time of researchers, purchasing a suite of commercial software or developing the in-house tools.

The type of data analysis software or tools that need to be used generally depends on the instrumental approach used to generate the data. For instance, data generated by NMR usually are processed by specific tools designed for aligning and annotating NMR spectra and are provided mainly by the instrument manufacturers [69]. Databases also are built based on type of samples analyzed to facilitate the process of identifying particular bin/s within NMR spectra [69,70]. Similarly, many software and tools also have been developed for the analysis of GC-MS and LC-MS data [71,72].

Open source software and web interfaces are now providing much better platforms for data analysis, starting from data mining to data interpretation [67,73]. Spicer, Salek, Moreno, Cañueto and Steinbeck [68] recently published a review article stating most of the freely available software tools for metabolomics data analysis. Their review covered tools that are used for data pre-processing, annotating, post-processing and statistical analysis, and readers are advised to consult that review to obtain a wide overview of the open source software [68]. Most of the available software tools are either R based [71,74–76] or Python based [72,77,78].

Metabolomics data analysis, like any other omics approaches, is moving towards cloud-based analysis. According to Warth, et al. [79], cloud computing provides multiple advantages over downloadable desktop based software mainly because of the straightforwardness of data sharing, transferring, managing and archiving. They also reinforced the fact that cloud computing allows a better standardization of data formatting and distribution in addition to ensuring global access of the data without the need for confined high-end computational hardware [79]. However, this process does not come without challenges as it requires a consistent and fast internet connection and often may face security issues in terms of intellectual property [79]. Regardless of the associated risks, cloud computing is becoming very popular for metabolomics data analysis and there are already platforms that make use of this approach, such as XCMS online [67] and Metaboanalyst [80]. Researchers across the different continents are now making use of these cloud-based metabolomics data analysis pipelines. For instance, Metaboanalyst was used by 60,000 researchers from 2000 cities around the world over the past 12 months, and approximately 6000 jobs per weekday or 150,000 jobs/month are usually submitted to this web interface [80]. XCMS online has over 4500 registered users from 120 different countries [67]. These data clearly indicates the popularity and usefulness of cloud-based metabolomics data analysis. The rise of cloud-based data analysis is not only allowing us to handling more data with ease, it is also helping us to tease out the biological meanings from the metabolomics data. However, we are still far from unravelling the true potential of all the datasets available within metabolomics community.

Continuous development of analytical instrumentations and data analysis platforms together is now providing us access to enormous amount of metabolomics data. This brings forward another important issue, and the metabolomics community is now discussing how to manage the openly available metabolomics data [81]. Community-based initiatives (e.g., MetaboLights and Metabolomics Workbench) develop tools that would allow the storing and exchanging the huge amount of heterogeneous data [82,83]. In addition, efforts of standardizing data sharing and reporting also began in 2007 by a metabolomics community driven initiative, the Metabolomics Standards Initiative (MSI) [84]. Recently, another initiative, COSMOS (Coordination of standards in metabolomics), has started its journey to fill the existing gaps in data reporting and sharing by taking some examples from other omics approaches [85]. Therefore, the metabolomics community is well aware of existing and upcoming challenges due to the omics revolution. This will in turn encourage the further improvements of data analysis pipelines.

3. Application of Metabolomics in Grape and Wine Research: State of the Art

The analysis of grape juice or wine samples can be problematic due to the complex matrix arising from either a high sugar or alcohol content. The detection of compounds present at very low concentrations in grape juice and wine can be hampered. Therefore, the matrix effect (ME) may result in poor and unreliable data as it has significant effect on the reproducibility, linearity and accuracy of the methods used by various analytical instruments [86]. ME is a key concern for the analysis of complex biological samples by LC-MS and many studies already have been undertaken to address this issue [87–93]. A sample clean-up step using SPE or SPME or liquid extraction is usually performed to avoid or reduce ME prior to analysis of samples by other methods [93]. More efficient chromatographic separation is also suggested by Trufelli, Palma, Famiglini and Cappiello [86]. However, these pre-analytical steps are time-consuming, arduous and often can cause loss of analytes, which is not appropriate for an unbiased profiling approach [86,87,94]. Details on sample preparation of grapes, wines and related microorganisms are provided in Lloyd, Johnson and Herderich [38].

Despite ME being a major issue for the analysis of grape juices and wines, comprehensive metabolite profiling is becoming an important tool these days. This approach has been efficaciously applied to distinguish white wines [95], to observe vintage effects on juices [96] and also to obtain information about grape chemical composition, wine typicity and quality [97]. Howell, et al. [98] also used metabolite profiling by GC-MS to determine the connections of different *Saccharomyces* species during wine fermentation. Comprehensive metabolome analysis of Sauvignon blanc grape juices and wines also revealed some new insights on the relationship of juice metabolites with key wine volatile metabolites [34]. Hence, untargeted metabolite profiling is indeed a favorable tool in grape and wine research. In addition, many researchers are using a targeted approach to determine specific groups of metabolites in grape juices, wines and wine yeasts [31,99,100]. However, it is noteworthy that even a targeted analysis using a high-resolution analytical platform is able to provide information of over hundreds of metabolites. For instance, lipidomics is one of the branch from targeted metabolomics that allows the determination of wide ranges of lipid species and fatty acids (often over 500) using a suitable analytical instrument [101]. On the other hand, an untargeted approach is gaining popularity for the analysis of volatile compounds in different biological samples, including wines, and often, this area is referred to as a "volatilome" [30,102–104]. Table 2 represents the comparisons between commonly used analytical instruments and their application in grape and wine research [13].

One of the major outcomes from metabolomics is the development of methods for the analysis of different groups of metabolites in complex grape and wine samples [33]. While targeted metabolomics led to the development of methods suitable for the analysis of particular group of metabolites with accuracy and sensitivity, untargeted metabolomics is now enabling us to detect thousands of metabolites just in a single run. Metabolomics has been applied to different areas of agriculture and food sciences albeit the number of publications is much lower compared to other areas (e.g., biomedical, cancer research). Based on the data obtained from Web of Science (on 4th October, 2018), a total of 198 and 154 articles (research and review) have been published in wine and grape metabolomics, respectively, within the last 13 years (Figure 2).

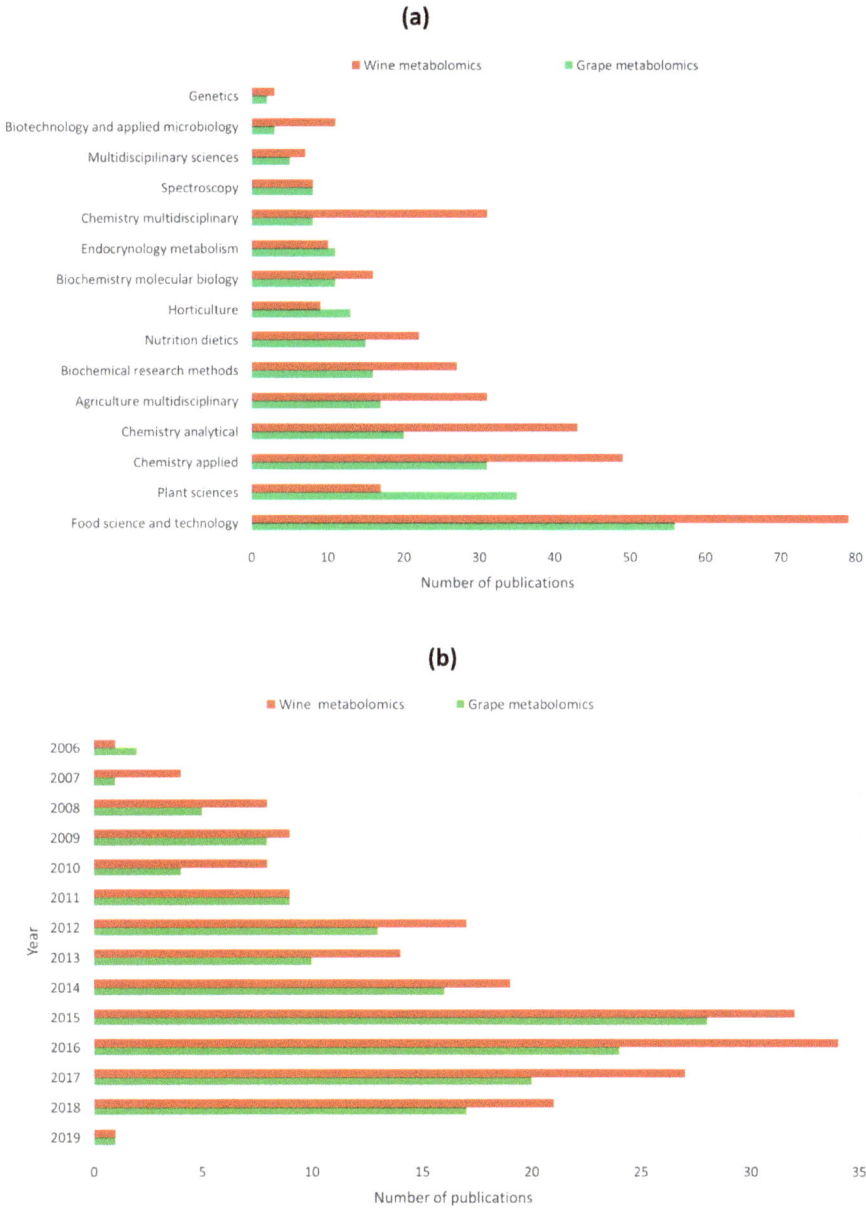

Figure 2. Publications on grape and wine metabolomics based on different research areas (**a**) and years (**b**). Data were obtained from the Web of Science.

Table 2. Comparisons among different analytical instruments used in targeted and untargeted metabolite analysis of grape juices and wines.

Analytical Technique	Advantages	Disadvantages	Use in Grape and Wine Research
GC-MS	• High chromatographic resolution • Sensitive and robust • Simultaneous analysis of different groups of metabolites • Large linear range, availability of commercial and in-house MS libraries	• Derivatization is required for non-volatile metabolites • Unable to analyze thermolabile compounds	[8,30,31,34,98–110]
LC-MS	• High sensitivity, Derivatization not usually required • Large sample capacity • Thermo-labile compounds can be analyzed	• Average to poor chromatographic resolution • De-salting may be required • Limited commercial libraries • Tough restrictions on LC eluents • Matrix effects	[36,111–119]
NMR	• Rapid analysis • Non-destructive • Minimal sample preparation • Quantitative	• Low sensitivity • More than one peak per component • Identification is laborious due to complex matrix	[59,60,95,107,120–123]
CE-MS	• High resolution, small volume of sample required • Rapid analysis • Usually no derivatization required	• Poor reproducibility • Poor sensitivity • Buffer incompatibility with MS • Difficulty in interfacing with MS • Limited commercial libraries	[124,125]

Interestingly, approximately 70% of these articles were published within last five years, which clearly shows that grape and wine researchers are more interested in adopting metabolomics either in a targeted or in an untargeted manner (Figure 2). Based on this published research, some of the most interesting applications are discussed below mainly by highlighting those published in last five years. Readers are also recommended to consult three other review articles where the potential applications of metabolomics in different aspects of grape and wine research have been discussed [35,37,38].

3.1. Untargeted Metabolomics As A Hypothesis-Generating Tool in Grape and Wine Science

The journey of metabolomics as an omics tool started over 20 years ago, mainly as a data- and question-driven approach. Therefore, most of the earlier publications in grape and wine metabolomics also aimed to determine as many metabolites as possible in order to develop some new insights into grape growing and winemaking [34,105,117,120]. In 2008, a technology feature was published in Nature that coined a word "wine-omics" to discuss a research project by Kirsten Skogerson at the University of California, Davis, where metabolomics was used as a data-driven approach [126]. Since then, many others adopted untargeted metabolomics to generate hypotheses regarding the role of wine yeasts and grape juice components in the development of wine aroma compounds. For instance, both GC-MS- and NMR-based methods were developed for the untargeted analysis of Sauvignon blanc grape juices and wines [34]. Combined data allowed the authors to generate some data-driven hypotheses for the role of different juice metabolites in the major wine aroma compounds (varietal thiols). The results from a simple juice manipulation experiment confirmed the hypotheses and showed the capability of metabolomics as a hypothesis-generating tool [34]. Similarly, Arapitsas, et al. [127] applied an untargeted LC-MS based metabolomics platform to study the role of micro-oxygenation during winemaking. They formulated hypotheses on the development and reactivity of wine pigment and the role of different primary and secondary metabolites in this matter, thus revealing the benefits of using unbiased, untargeted metabolomics to advance their understanding of wine chemistry [127].

Since the beginning, there was an ongoing need to develop improved analytical platforms that would allow the determination of a large number of metabolites within the grape and wine metabolome. The use of high-resolution MS instrumentations, such as, fourier transform mass spectrometry (FTICR-MS) and ultra-high performance liquid chromatography coupled with quadrupole time-of-flight mass spectrometry (UPLC-Q-ToF-MS) in wine analysis in an untargeted manner is now making it possible to detect metabolites with high precision and mass accuracy [32,116–118,128]. In addition to revealing the true complexity of the wines, this approach is now adding another dimension in terms of obtaining exact mass for formula calculation with retention time information of unknown molecules [117]. Liu, Forcisi, Harir, Deleris-Bou, Krieger-Weber, Lucio, Longin, Degueurce, Gougeon, Schmitt-Kopplin and Alexandre [128] also applied untargeted metabolite profiling using FTICR-MS and UPLC-Q-ToF-MS to determine the outcomes from the interaction of malolactic bacteria and yeasts that either stimulate (MLF+) or inhibit (MLF−) malolactic fermentation. In this study, they were able to detect 3000 discriminant masses that characterized the phenotypes of both MLF+ and MLF− yeast strains in addition to determining MLF− biomarkers. A combination of both targeted and untargeted metabolomics approaches was also found to be beneficial in determining the role of ethanol stress in an off-odor producing yeast, *Dekkera bruxellensis*, and also in generating new knowledge on this contaminant yeast [129].

An untargeted metabolomics approach usually provides an opportunity to look at the system in a holistic way and encourages thinking outside the box and not to be reductive. Therefore, the application of this approach generated a mammoth amount of data in grape and wine research [33,34,109,130]. In the near future and with the development of suitable data analysis platforms, these data can be explored to their full potential. Thus, it shows the promise of new innovation and generation of new knowledge to fill the current gaps.

3.2. Study of Terroir, Authenticity and Originality of Grapes and Wines Using a Metabolomics Approach

Wine is a comparatively expensive commodity in modern society, and winemaking is considered not only a science, but also an art. Therefore, both originality and *terroir* are important aspects for the wine producers, particularly for wine makers from old world countries (e.g., European countries). Metabolomics approaches have been applied in this area of wine science to provide analytical tools that would allow differentiation among different wine growing regions [59,118,131], quality control and authentication of wines [55,132]. In a review article published in 2015, Alanon, Perez-Coello and Marina [35] provided a comprehensive discussion on the application of metabolomics by using different instrumentation platforms on wine traceability. Therefore, this section will mostly highlight the works in this subject matter published from 2014 until now.

NMR was the instrument of choice for most of the studies that dealt with *terroir*, authenticity and originality of grapes and wines [58,60,123,131] and Amargianitaki and Spyros [132] provided an excellent overview of application of NMR-based metabolomics in this area. Some of the most recent studies [60,133] that made use of NMR-based metabolomics again re-inforce the fact that NMR is a powerful instrumental approach with high reproducibility and requires minimum sample preparation [55]. For instance, Cassino, Tsolakis, Bonello, Gianotti and Osella [133] applied ^{1}H NMR-based metabolomics and chemometrics to differentiate the grapes produced within the Barbera regions of Italy. In addition, they also determined the influence of different climatic factors on the wine composition. One of the interesting studies from Picone, Trimigno, Tessarin, Donnini, Rombola and Capozzi [60] reported the differences among grapes produced by different cultivation systems (biodynamic and organic) using comprehensive NMR analysis. They found a lower amount of sugars, coumaric and caffeic acids and higher concentrations of proline, valine and γ-aminobutyric acid (GABA) in biodynamic grapes than in organic ones. These results clearly indicate that cultivation practices alter the grape metabolome and as a result will also have significant effect on wine quality. Although NMR is widely used for grape and wine analysis, it is noteworthy that this instrument is mainly capable of detecting metabolites that are usually present in higher concentrations [34].

In comparison to NMR, MS provides far better coverage of metabolites present in any biological samples and many studies have been published last five years to demonstrate the application of MS-based metabolomics in determining growing regions of grapes and wines. For example, Roullier-Gall, et al. [134,135] developed analytical platforms using high-resolution MS techniques that were able to detect over thousands of features in grape and respective wine samples. Using these data, they showed the effects of geographical location and vintages on the grape and wine composition produced in Burgundy regions. Bokulich, et al. [136] reported another interesting study demonstrating the relationship between berry microbiome and metabolome and their combined effect on wine *terroir*. They surveyed over 200 commercial wine fermentations within Napa and Sonoma wine counties and determined the wine metabolite profiles using UHPLC-QTOF-MS. Using machine learning models, they showed that the bacterial and fungal consortia in wines correlate with the chemical composition of the finished wines, thus directly influencing the regional characteristics of the wines [136].

A combination of different MS-based metabolomics approaches also seemed to be successful in finding out the terroir effect on grapes and wines. Anesi, et al. [137] applied both GC-MS- and LC-MS-based metabolomics to analyze grape berries from a single clone of the Corvina variety grown in seven different vineyards, located in three macrozones, over a 3-year trial period to determine the effect of *terroir*. Their results showed that the berry metabolome is mainly affected by the vintage. While some of the non-volatile (e.g., stilbene, anthocyanins and flavonoids) and volatile metabolites showed a trend of plasticity over the three vintages, other metabolites including procyanindins and flavan-3-ols seemed to be much more stable.

3.3. Study of Yeast Metabolism and Aroma Compound Development during Wine Making

The use of both targeted and untargeted metabolite analysis provides a snapshot of any microbial metabolism based on the growing environment [108,138]. The fermentation process during winemaking is mainly dominated either by a single inoculated commercial wine yeast strain or by a number of wine yeasts already present in the grape juices [139]. Regardless of the type of fermentation, the environmental condition is generally not favorable for any types of yeasts considering grape juice is a high sugar (hyperosmotic stress) and low nitrogen (nutrient limited) growth medium [140] (Figure 3). Moreover, once the fermentation begins and ethanol is produced, wine yeasts go through oxidative stress and ethanol toxicity [108] (Figure 3). As metabolomics allows the analysis of hundreds of metabolites in a single run, it is an excellent tool for the study of the metabolic behavior of wine yeast strains.

Over the past decade, many metabolomics studies have been carried out to determine effect of juice or growth media composition on overall wine yeast metabolism, with particular attention on the developments of fermentation end products [107,108,141]. For instance, two publications from a research group of the University of Auckland reported changes before and after fermentation in Sauvignon blanc juices and how juice composition influences the major varietal aroma compound production by *Saccharomyces cerevisiae* EC1118 [34,107]. Based on a combination of comprehensive metabolite analysis by two different analytical platforms (GC-MS and NMR), they analyzed 63 grape juices and respective wines produced over six different seasons and showed that assimilation of different nitrogen and carbon sources by EC1118 depended on the overall grape juice composition [107]. Moreover, their studies provided some new insights into the metabolism of a wine yeast strain and generated new hypotheses about the potential roles of juice metabolites on the development of varietal aroma compounds. Their data together with other published information provided knowledge on how the biosynthetic pathways of secondary metabolites (e.g., thiols) are expected to be highly interconnected to primary central carbon metabolic pathways. Therefore, any alterations in one or more of these primary metabolic pathways are likely to influence the biosynthetic tariffs of secondary metabolites (Figure 3).

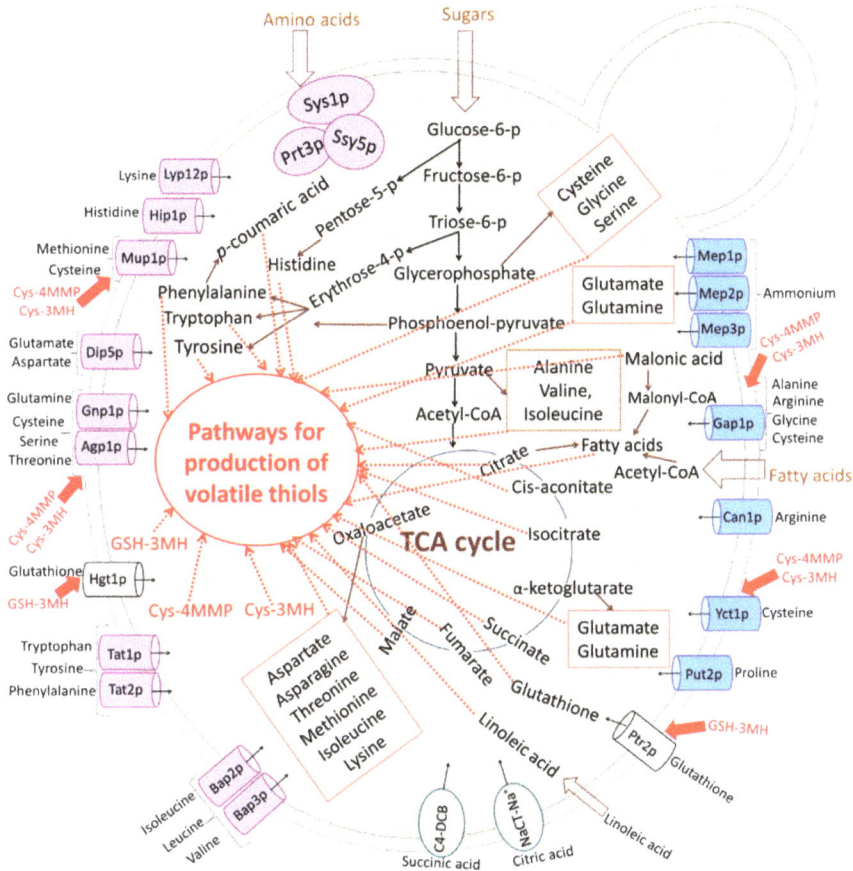

Figure 3. The anticipated metabolic network related to varietal thiol pathways in *Saccharomyces cerevisiae* based on the metabolomics data published by Pinu, Edwards, Jouanneau, Kilmartin, Gardner and Villas-Boas [34]. This figure shows the transport mechanisms of different metabolites into the yeast cells. The amino acid transporters regulated by Ssy-Ptr3-Ssy5 (SPS) genes are shown in purple and the transporters regulated by nitrogen catabolite repression (NCR) are shown in blue. It also presents the major pathways of yeast metabolism and how they are connected to the formation of volatile thiols. The red (solid) arrows indicate the suggested ways for the transportation of putative precursors to the yeast cells, while the dashed red arrows indicate which metabolites (e.g., nitrogenous compounds, organic acids and fatty acids) influence the production of varietal thiols in Sauvignon blanc wines. The information for this figure was collated from Pinu [13], Aliverdieva, et al. [142], Crépin, et al. [143], Cooper [144], Hofman-Bang [145], Henscke and Jiranek [146]. Here, cys-3MH and cys-4MMP denote the cysteinylated precursors of 3-mercaptohexanol (3MH) and 4-mercapto-4-methylpentan-2-ol, respectively, and GSH-3MH is the glutathionylated precursor of 3MH.

The influence of different groups of metabolites, particularly amino and fatty acids, and fermentation conditions on winemaking and also on wine yeast metabolism is a well-studied area [108,141,147–151]. A considerable amount of new knowledge has been generated in this matter mostly by using targeted metabolomics or a metabolite-analysis approach. Most of these studies were performed either using a natural grape juice medium [108,141] or grape juice-like synthetic media [149,150]. Casu, Pinu, Fedrizzi, Greenwood and Villas-Boas [108] investigated the effect of pre-fermentative supplementation of an unsaturated fatty acid, linoleic acid, on the metabolism and aroma compound production by three different wine yeast strains. Using a GC-MS-based metabolomics approach, they showed that increased concentration of linoleic acid affected different primary (e.g., amino acids) and secondary metabolites (e.g., varietal thiols and acetate esters) in a strain-specific manner. Another experiment from the same laboratory also revealed that pre-supplementation of different saturated and unsaturated fatty acids during Sauvignon blanc fermentation significantly affected the metabolism of wine yeast and as a result, end product formation [141]. These metabolomics studies together generated an awareness for the wider wine industry to select wine yeast based on the juice composition in order to obtain wines with desired aromatic profiles.

The effect of temperature and micro oxygenation/oxygen impulse on wine yeast metabolism is another area of wine science where metabolomics was used as a tool [150,152]. For example, López-Malo, Querol and Guillamon [150] compared the metabolome of commercial wine yeast strains (*S. cerevisiae*, *S. bayanus var. uvarum* and *S. kudriavzevii*) at 12 °C and 28 °C in a synthetic grape must medium. Their data confirmed that cryotolerant yeast species (*S. bayanus var. uvarum* and *S. kudriavzevii*) responded differently to the temperature difference than the *S. cerevisiae* strain and the main difference was observed in carbohydrate metabolism. Moreover, an elevated shikimate pathway activity was found in *S. bayanus var. uvarum*, while NAD^+ synthesis increased in *S. kudriavzevii* in response to cold temperature. Another study from Rollero, Bloem, Camarasa, Sanchez, Ortiz-Julien, Sablayrolles, Dequin and Mouret [152] reported the development of a model to predict the combined effect of nutrition (nitrogen and lipid) and temperature on the production of fermentative aroma compounds by *S. cerevisiae* EC1118 during winemaking. Their results again proved the complex effect of different environmental parameters on non-volatile and volatile metabolites and shed new light on the synthesis and regulation of secondary metabolites.

3.4. Combination of Metabolomics and Transcriptomics to Unravel New Knowledge

As a metabolomics approach provides information on the most downstream products of a cell system, the combination of this tool with other omics approaches can be extremely powerful and may allow the generation of unique knowledge or help to fill the knowledge gaps [139]. Many multi-omics integration studies have already been performed to elucidate more details of *S cerevisiae* metabolism as a cell factory. This integrated approach is still under-utilized in grape and wine research compared to other food and agricultural sectors and only few publications can be found. However, a combination of metabolomics and transcriptomics has been adopted to study different wine yeast strains under wine fermentation-like conditions to develop new insights into the role of oxygen impulse on *S cerevisiae* wine strains [153,154] and to explore the aroma profiles of wines produced by different *S cerevisiae* strains of diverse origins [155].

Stuck fermentation can be a huge problem in winemaking and various technologies are already in use to re-start the fermentation process. Among those, oxygen impulse is often used to avoid the loss of a whole batch of wines. Using integrated metabolomics and transcriptomics, Aceituno, Orellana, Torres, Mendoza, Slater, Melo and Agosin [153] investigated the effects of different ranges of dissolved oxygen on the metabolism of *S cerevisiae* EC1118 grown under a carbon-sufficient but nitrogen-limited medium. They showed that an increase in dissolved oxygen from 1.2 to 2.7 μM caused the yeast cells to change their metabolism from a fermentative to a mixed respiro-fermentative one, which was characterized by a shift in the process of the tricarboxylic acid cycle (TCA) and an activation of NADH transferring from the cytosol to mitochondria. They also observed a significant change in several key respiratory genes, and also in genes related to proline uptake, cell wall remodelling, and oxidative stress. In addition, their results indicated that respiration was accountable for a large portion of the oxygen response in yeast cells during alcoholic fermentation. Another study from the same group also showed the physiological response of *S cerevisiae* EC1118 strain to the sudden increase of dissolved oxygen in a carbon-sufficient but nitrogen-limited medium [154]. Their results confirmed the induction of genes related to mitochondrial respiration, ergosterol biosynthesis, and oxidative stress and the repression of mannoprotein coding genes in response to the increased amount of dissolved oxygen in the growth medium. However, the authors highlighted the fact that oxygen plays a dual role in winemaking considering some potential detrimental effects on wine aroma, although their integrated omics approach provided some new knowledge on the beneficial influence of oxygen availability on wine yeast metabolism [154].

Integration of transcriptomics and metabolomics data can also be used for the generation of new insights into the yeast metabolism and fermentation process, particularly aroma compound development. For instance, Mendes, Sanchez, Franco-Duarte, Camarasa, Schuller, Dequin and Sousa [155] performed a comparative transcriptomics and metabolomics analysis of four yeast strains from diverse origins (e.g., wine, sake, cacha double dagger and laboratory) at two time points. They used a multivariate factorial analysis to identify new markers that can be used for improvement of aroma production. The combined omics data allowed them to differentiate all the yeast strains at both the metabolic and transcriptomic level and extended their knowledge on the production of wine aroma and flavor. They also identified new genes associated with the development of flavor active compounds, primarily those related to the production of fatty acids, and ethyl and acetate esters [155].

3.5. Application of Metabolomics to Study Grape Growth Developments and Grape Vine Disease

Metabolomics is comparatively a new tool for viticultural studies and only a few publications are available that show the suitability of this omics approach, mostly in a targeted manner to study the vine and grape berry development [59,110,156,157]. Recently, Cuadros-Inostroza, Ruíz-Lara, González, Eckardt, Willmitzer and Peña-Cortés [110] published a GC-MS based metabolomics study that aimed at gathering more information on primary metabolites during the grape berry development in different cultivars. They analyzed grape berry samples of two cultivars across six stages (from flowering to maturity) using an untargeted metabolite profiling approach. They identified 115 metabolites in those samples and showed that the changes in metabolite composition were growth stage specific and particularly more distinct during fruit setting and pre-veraison. Moreover, they performed a network analysis and confirmed again that network connectivity of primary metabolites was stage and cultivar dependent. Therefore, they suggested some association between primary metabolites during berry developmental processes between different grapevine cultivars [110]. Another study from Hochberg, Degu, Cramer, Rachmilevitch and Fait [156] also determined the differences in berry metabolism in Shiraz and Cabernet Sauvignon vines under shortfall of irrigation by determining berry skin metabolite profiles using both GC-MS and LC-MS. They clearly observed different types of effect of water deficit on berry metabolism between these two cultivars and showed that water deficit increased the production of stress-related metabolites (e.g., proline, beta-alanine, nicotinate, raffinose and ascorbate) more in Shiraz than the Cabernet Sauvignon grapes. Moreover, water stress affected the polyphenol metabolism uniquely for each cultivar. This metabolomics study revealed a link between the vine hydraulics and water stress-related differences in berry skin metabolism [156].

In addition to grape berry development, early detection of grape vine disease through metabolomics may become an important area of research. A recent study reported some interesting findings on a grape vine disease (grape vine downy mildew) pathosystem using a multi-omics (genomics, transcriptomics and metabolomics) approach [158]. The combination of multiple omics data allowed the authors to characterize the pathosystem of downy mildew causing pathogen, *Plasmopara viticola*, in a molecular and biochemical level. They also determined a potential RNA-based marker that can be used to screen novel resistant grape varieties. Other earlier work by Benheim, et al. [159] also showed the capability of the metabolomics approach in determining potential biomarkers for an early detection method for phylloxera infestation in grapevine. They identified four flavonoid compounds: isorhamnetin glycoside, rutin, kaempherol glycoside and quercetin glycoside in grape vine leaf using LC-MS and validated the specificity of these compounds against phylloxera through field and glasshouse-based trials against nutrient and water stress. A further NMR based metabolomics study of grape vine leaves from phylloxera infested and uninfested vines also revealed the key metabolic changes that occur in the grape vine [160]. For instance, they found that sucrose, caffeic acid and quercetin were up-regulated in infested grapevines while glucose and leucine were downregulated. Based on both of these studies, they suggested quercetin as a potential marker that could be used for the early detection of phylloxera infestation in grape vines.

4. Future Perspectives

In comparison to classical chemical analysis of grape juices and wines, targeted and untargeted metabolomics approaches are more beneficial as these approaches are able to cover more metabolites, including those present in very small concentrations. Both analytical platforms and data analysis pipelines have improved considerably; therefore, our knowledge on different types of metabolites and related pathways has also advanced considerably compared with the last decade. The application of untargeted metabolomics in grape and wine research resulted in the generation of huge amount of data. However, it is still unfortunate that we are currently unable to identify many of the metabolites present within the grape and wine metabolome. Without identification, the interpretation of the metabolomics data is not complete and sometimes meaningless as it is impossible to connect metabolites without identity with pathways [161]. This somehow caused confusions and many are adopting targeted metabolome analysis of specific groups of metabolites within grape and wine research. With targeted analysis, it is now possible to detect and identify over hundreds of metabolites in a single run, particularly using MS-based analytical techniques.

Further developments of analytical instrumentations and data analysis platforms within the next years will allow the generation of even larger data sets. However, the metabolomics community is now well aware of the upcoming issue of omics data revolution and many research groups are now building platforms to make metabolomics data more accessible and manageable [81]. There is also a growing interest in making metabolite identification more robust [161]. This will indeed support the notion of making more sense out of the data obtained from any metabolomics experiment. For instance, machine learning and deep learning approaches are improving; this will in turn be useful to tease out those ever increasing data sets by determining the key features, thus allowing better interpretation of metabolomics and other omics data [162,163].

Efforts have already been made to create databases of metabolites present within a particular system (e.g., food metabolome) [164]. Similarly, grape and wine metabolite databases will also be publicly available, thus providing more information on how grape juice and respective wine composition varies from season to season and region to region [33]. For example, in New Zealand, a comprehensive research program on Sauvignon blanc grape juices and wines has been conducted with the help of the government, and industry provided funding that built a nationwide collaboration among universities and research institutes. Under this program, a database has been created that contains metabolite and other compositional data sets of over 400 grape juices and wines collected over three different seasons and from all the wine-producing regions. This database also harbors information provided by the grape growers and weather stations. Further work on building a predictive tool for the winemakers is ongoing by using these data sets (data not yet published).

It is clear that the application of single omics is still unable to provide a holistic overview of any biological system. Therefore, combinations or the integration of omics methods is becoming increasingly popular although there are still many limitations. The future of grape and wine research will also benefit from the adoption of multi-omics or system biology approaches [140]. Metabolomics combined with genomics, proteomics and transcriptomics could be an extremely powerful tool to study wine yeast metabolism and the overall fermentation process. Connection of metabolites with related genes, proteins and RNAs could lead to the generation of new knowledge on metabolism of wine yeast and other related microorganisms in addition to discovering or improving different microbial strains via metabolic engineering to be used during fermentation.

The field of genome scale modelling is substantially developed [165], and this platform can be used to connect genes, proteins and related metabolites. A genome-scale metabolic model

(GEM) or reconstruction of different industrially significant microbes (e.g., *S cerevisiae*) is already available [166,167]. Over 12 genome-scale metabolic reconstructions are available for *S cerevisiae* [168]. Professor Eduardo Agosin's research group from Pontifical Catholic University of Chile has done a considerable amount of work on creating GEMs for different wine microorganisms including *Oenococcus oeni* and *Pichia pastoris* [169–172]. Therefore, guidelines are already available and this approach can be easily applied to microbial strains used in winemaking. Predictive tools for the winemakers can be developed by integrating metabolic modelling and metabolomics approaches. The wine research community and wine industry will benefit if such tools could be available to aid decision making on what type of juice or yeast strain should be used for the production of wine styles based on consumers' demand.

5. Conclusions

Due to the availability of analytical instruments with high resolution and exceptional sensitivity, the analysis of metabolites either in an untargeted or a targeted manner allows us to determine a large number of metabolites. Data analysis pipelines have also developed significantly; therefore, application of metabolomics in grape and wine research is increasing exponentially. Although the ever-increasing number of unknown metabolites is hindering the data interpretation process, this can be overcome in the near future, and community efforts are already underway to deal with this particular issue. As mentioned earlier, the application of metabolomics in grape and wine research is more recent and many are exploring the potential of this approach. However, metabolomics combined with other omics approaches is becoming extremely useful and can be applied in all sectors of grape and wine research. Thus far, we have already generated a significant amount of data and knowledge on grape and wine production systems. In future, this area will most probably lead towards the development of approaches to combine genome-scale metabolic modelling with metabolomics and/or integration of multiomics.

funding: This research received no external funding.

Conflicts of Interest: The author declares no conflict of interest.

References

1. Webb, A.D. Applications of gas chromatography in studying the aromatic qualities of wines. *Qual. Plant. Mater. Veg.* **1964**, *11*, 234–243. [CrossRef]
2. Noble, A.C.; Flath, R.A.; Forrey, R.R. Wine headspace analysis. Reproducibility and application to varietal classification. *J. Agric. Food Chem.* **1980**, *28*, 346–353. [CrossRef]
3. Guth, H. Quantitation and sensory studies of character impact odorants of different white wine varieties. *J. Agric. Food Chem.* **1997**, *45*, 3027–3032. [CrossRef]
4. Mestres, M.; Busto, O.; Guasch, J. Headspace solid-phase microextraction analysis of volatile sulphides and disulphides in wine aroma. *J. Chromatogr. A* **1998**, *808*, 211–218. [CrossRef]
5. Joslyn, M.A.; Dunn, R. Acid metabolism of wine yeast. I. The relation of volatile acid formation to alcoholic fermentation. *J. Am. Chem. Soc.* **1938**, *60*, 1137–1141. [CrossRef]
6. Nelson, R.R.; Acree, T.E.; Butts, R.M. Isolation and identification of volatiles Catawba wine. *J. Agric. Food Chem.* **1978**, *26*, 1188–1190. [CrossRef]
7. Cobb, C.S.; Bursey, M.M. Comparison of extracting solvents for typical volatile components of eastern wines in model aqueous-alcoholic systems. *J. Agric. Food Chem.* **1978**, *26*, 197–199. [CrossRef]
8. Kwan, W.O.; Kowalski, B.R. Pattern recognition analysis of gas chromatographic data. Geographic classification of wines of *Vitis vinifera* cv. Pinot noir from France and the united states. *J. Agric. Food Chem.* **1980**, *28*, 356–359. [CrossRef]
9. Mayer, K.; Busch, I. On th enzymatic determination of glyerin in the grape juice and wine. *UMitteilungen Gebiete Lebensmittel Hyg.* **1963**, *54*, 297–303.
10. Caldwell, J.S. Some effects of seasonal conditions upon the chemical composition of American grape juice. *J. Agric. Res.* **1925**, *30*, 1133–1176.

11. Webster, J.E.; Cross, F.B. *Chemical Analysis of Grape Juices: Varietal Comparisons*; American Society of Horticultural Sciences: Alexandria, VA, USA, 1936.

12. Lafon-Lafourcade, S. Enzymatic methods in the analysis of musts and wines. *Appl. Méthodes Enzymatiques L'anal. Mouts Vins* **1978**, *32*, 969–974.

13. Pinu, F.R. *Sauvignon Blanc Metabolomics: Metabolite Profile Analysis before and after Fermentation*; University of Auckland: Auckland, New Zealand, 2013.

14. Oliver, S.G. Systematic functional analysis of the yeast genome. *Trends Biotechnol.* **1998**, *16*, 373–378. [CrossRef]

15. Villas-Bôas, S.G.; Rasmussen, S.; Lane, G.A. Metabolomics or metabolite profiles? *Trends Biotechnol.* **2005**, *23*, 385–386. [CrossRef] [PubMed]

16. Patti, G.J.; Yanes, O.; Siuzdak, G. Metabolomics: The apogee of the omics trilogy. *Nat. Rev. Mol. Cell Biol.* **2012**, *13*, 263–269. [CrossRef] [PubMed]

17. Creydt, M.; Fischer, M. Plant metabolomics: Maximizing metabolome coverage by optimizing mobile phase additives for nontargeted mass spectrometry in positive and negative electrospray ionization mode. *Anal. Chem.* **2017**, *89*, 10474–10486. [CrossRef] [PubMed]

18. Jacob, M.; Malkawi, A.; Albast, N.; Al Bougha, S.; Lopata, A.; Dasouki, M.; Rahman, A.M.A. A targeted metabolomics approach for clinical diagnosis of inborn errors of metabolism. *Anal. Chim. Acta* **2018**, *1025*, 141–153. [CrossRef] [PubMed]

19. Dervishi, E.; Zhang, G.; Mandal, R.; Wishart, D.S.; Ametaj, B.N. Targeted metabolomics: New insights into pathobiology of retained placenta in dairy cows and potential risk biomarkers. *Animal* **2018**, *12*, 1050–1059. [CrossRef] [PubMed]

20. Fiehn, O.; Kopka, J.; Dörmann, P.; Altmann, T.; Trethewey, R.N.; Willmitzer, L. Metabolite profiling for plant functional genomics. *Nat. Biotechnol.* **2000**, *18*, 1157–1161. [CrossRef] [PubMed]

21. Nielsen, J.; Jewett, M.C. The role of metabolomics in systems biology. In *Topics in Current Genetics*; Springer: Berlin/Heidelberg, Germany, 2007; Volume 18, pp. 1–10.

22. Sumner, L.W.; Mendes, P.; Dixon, R.A. Plant metabolomics: Large-scale phytochemistry in the functional genomics era. *PhytoChem* **2003**, *62*, 817–836. [CrossRef]

23. Trethewey, R.N. Gene discovery via metabolic profiling. *Cur. Opin. Biotechnol.* **2001**, *12*, 135–138. [CrossRef]

24. Bino, R.J.; Hall, R.D.; Fiehn, O.; Kopka, J.; Saito, K.; Draper, J.; Nikolau, B.J.; Mendes, P.; Roessner-Tunali, U.; Beale, M.H.; et al. Potential of metabolomics as a functional genomics tool. *Trends Plant Sci.* **2004**, *9*, 418–425. [CrossRef] [PubMed]

25. Dyar, K.A.; Eckel-Mahan, K.L. Circadian metabolomics in time and space. *Front. Neurosci.* **2017**, *11*, 369. [CrossRef] [PubMed]

26. Saito, K.; Matsuda, F. Metabolomics for Functional Genomics, Systems Biology, and Biotechnology. *Annu. Rev. Plant Biol.* **2010**, *61*, 463–489. [CrossRef] [PubMed]

27. Dixon, R.A.; Gang, D.R.; Charlton, A.J.; Fiehn, O.; Kuiper, H.A.; Reynolds, T.L.; Tjeerdema, R.S.; Jeffery, E.H.; German, J.B.; Ridley, W.P.; et al. Perspective—Applications of metabolomics in agriculture. *J. Agric. Food Chem.* **2006**, *54*, 8984–8994. [CrossRef] [PubMed]

28. Wishart, D.S. Metabolomics: Applications to food science and nutrition research. *Trends Food Sci. Technol.* **2008**, *19*, 482–493. [CrossRef]

29. Wang, D.D.; Zhang, L.X.; Huang, X.R.; Wang, X.; Yang, R.N.; Mao, J.; Wang, X.F.; Wang, X.P.; Zhang, Q.; Li, P.W. Identification of nutritional components in black sesame determined by widely targeted metabolomics and traditional Chinese medicines. *Molecules* **2018**, *23*, 12. [CrossRef] [PubMed]

30. Beckner Whitener, M.E.; Stanstrup, J.; Panzeri, V.; Carlin, S.; Divol, B.; Du Toit, M.; Vrhovsek, U. Untangling the wine metabolome by combining untargeted SPME–GCXGC-TOF-MS and sensory analysis to profile sauvignon blanc co-fermented with seven different yeasts. *Metabolomics* **2016**, *12*, 53. [CrossRef]

31. Martins, C.; Brandao, T.; Almeida, A.; Rocha, S.M. Metabolomics strategy for the mapping of volatile exometabolome from saccharomyces spp. Widely used in the food industry based on comprehensive two-dimensional gas chromatography. *J. Sep. Sci.* **2017**, *40*, 2228–2237. [CrossRef] [PubMed]

32. Billet, K.; Houille, B.; de Bernonville, T.D.; Besseau, S.; Oudin, A.; Courdavault, V.; Delanoue, G.; Guerin, L.; Clastre, M.; Giglioli-Guivarc'h, N.; et al. Field-based metabolomics of vitis vinifera l. Stems provides new insights for genotype discrimination and polyphenol metabolism structuring. *Front. Plant Sci.* **2018**, *9*, 15. [CrossRef] [PubMed]

33. Flamini, R.; De Rosso, M.; De Marchi, F.; Dalla Vedova, A.; Panighel, A.; Gardiman, M.; Bavaresco, L. Study of grape metabolomics by suspect screening analysis. In *IX International Symposium on Grapevine Physiology and Biotechnology*; Pinto, M., Ed.; International Society for Horticultural Science: Leuven, Belgium, 2017; Volume 1157, pp. 329–335.

34. Pinu, F.R.; Edwards, P.J.B.; Jouanneau, S.; Kilmartin, P.A.; Gardner, R.C.; Villas-Boas, S.G. Sauvignon blanc metabolomics: Grape juice metabolites affecting the development of varietal thiols and other aroma compounds in wines. *Metabolomics* **2014**, *10*, 556–573. [CrossRef]

35. Alanon, M.E.; Perez-Coello, M.S.; Marina, M.L. Wine science in the metabolomics era. *Trac-Trends Anal. Chem.* **2015**, *74*, 1–20. [CrossRef]

36. Arapitsas, P.; Ugliano, M.; Perenzoni, D.; Angeli, A.; Pangrazzi, P.; Mattivi, F. Wine metabolomics reveals new sulfonated products in bottled white wines, promoted by small amounts of oxygen. *J. Chromatogr. A* **2016**, *1429*, 155–165. [CrossRef] [PubMed]

37. Cozzolino, D. Metabolomics in grape and wine: Definition, current status and future prospects. *Food Anal. Meth.* **2016**, *9*, 2986–2997. [CrossRef]

38. Lloyd, N.; Johnson, D.L.; Herderich, M.J. Metabolomics approaches for resolving and harnessing chemical diversity in grapes, yeast and wine. *Aust. J. Grape Wine Res.* **2015**, *21*, 723–740. [CrossRef]

39. Roullier-Gall, C.; Heinzmann, S.S.; Garcia, J.-P.; Schmitt-Kopplin, P.; Gougeon, R.D. Chemical messages from an ancient buried bottle: Metabolomics for wine archeochemistry. *NPJ Sci. Food* **2017**, *1*. [CrossRef]

40. Hall, R.; Beale, M.; Fiehn, O.; Hardy, N.; Sumner, L.; Bino, R. Plant metabolomics: The missing link in functional genomics strategies. *Plant Cell* **2002**, *14*, 1437–1440. [CrossRef] [PubMed]

41. Wishart, D.S.; Knox, C.; Guo, A.C.; Eisner, R.; Young, N.; Gautam, B.; Hau, D.D.; Psychogios, N.; Dong, E.; Bouatra, S.; et al. Hmdb: A knowledgebase for the human metabolome. *Nucleic Acids Res.* **2009**, *37*, D603–D610. [CrossRef] [PubMed]

42. Stitt, M.; Sulpice, R.; Keurentjes, J. Metabolic networks: How to identify key components in the regulation of metabolism and growth. *Plant Physiol.* **2010**, *152*, 428–444. [CrossRef] [PubMed]

43. Villas-Bôas, S.G.; Mas, S.; Åkesson, M.; Smedsgaard, J.; Nielsen, J. Mass spectrometry in metabolome analysis. *Mass Spectrom. Rev.* **2005**, *24*, 613–646. [CrossRef] [PubMed]

44. Shulaev, V. Metabolomics technology and bioinformatics. *Brief. Bioinform.* **2006**, *7*, 128–139. [CrossRef] [PubMed]

45. Goeddel, L.C.; Patti, G.J. Maximizing the value of metabolomic data. *Bioanalysis* **2012**, *4*, 2199–2201. [CrossRef] [PubMed]

46. Han, J.; Datla, R.; Chan, S.; Borchers, C.H. Mass spectrometry-based technologies for high-throughput metabolomics. *Bioanalysis* **2009**, *1*, 1665–1684. [CrossRef] [PubMed]

47. Mas, S.; Villas-Bôas, S.G.; Hansen, M.E.; Åkesson, M.; Nielsen, J. A comparison of direct infusion MSand GC-MS for metabolic footprinting of yeast mutants. *Biotechnol. Bioeng.* **2007**, *96*, 1014–1022. [CrossRef] [PubMed]

48. Glinski, M.; Weckwerth, W. The role of mass spectrometry in plant systems biology. *Mass Spectrom. Rev.* **2006**, *25*, 173–214. [CrossRef] [PubMed]

49. Pope, G.A.; MacKenzie, D.A.; Defernez, M.; Aroso, M.A.M.M.; Fuller, L.J.; Mellon, F.A.; Dunn, W.B.; Brown, M.; Goodacre, R.; Kell, D.B.; et al. Metabolic footprinting as a tool for discriminating between brewing yeasts. *Yeast* **2007**, *24*, 667–679. [CrossRef] [PubMed]

50. Dunn, W.B. Mass Spectrometry in Systems Biology: An Introduction. In *Methods in Enzymology*; Daniel, J., Malkhey, V., Hans, V.W., Eds.; Academic Press: Oxford, UK, 2011; Volume 500, pp. 15–35.

51. Huo, Y.Q.; Kamal, G.M.; Wang, J.; Liu, H.L.; Zhang, G.N.; Hu, Z.Y.; Anwar, F.; Du, H.Y. H-1 NMR-based metabolomics for discrimination of rice from different geographical origins of China. *J. Cereal Sci.* **2017**, *76*, 243–252. [CrossRef]

52. Li, Q.Q.; Yu, Z.B.; Zhu, D.; Meng, X.H.; Pang, X.M.; Liu, Y.; Frew, R.; Chen, H.; Chen, G. The application of NMR-based milk metabolite analysis in milk authenticity identification. *J. Sci. Food Agric.* **2017**, *97*, 2875–2882. [CrossRef] [PubMed]

53. Fotakis, C.; Kokkotou, K.; Zoumpoulakis, P.; Zervou, M. NMR metabolite fingerprinting in grape derived products: An overview. *Food Res. Int.* **2013**, *54*, 1184–1194. [CrossRef]

54. Lee, J.E.; Lee, B.J.; Chung, J.O.; Hwang, J.A.; Lee, S.J.; Lee, C.H.; Hong, Y.S. Geographical and climatic dependencies of green tea (*Camellia sinensis*) metabolites: A H-1 NMRr-based metabolomics study. *J. Agric. Food Chem.* **2010**, *58*, 10582–10589. [CrossRef] [PubMed]

55. Hong, Y.S. NMR-based metabolomics in wine science. *Magn. Reson. Chem.* **2011**, *49*, S13–S21. [CrossRef] [PubMed]

56. Dieterle, F.; Riefke, B.; Schlotterbeck, G.; Ross, A.; Senn, H.; Amberg, A. NMR and MS methods for metabonomics. *Methods Mol. Biol.* **2011**, *691*, 385–415. [PubMed]

57. Alves, E.G.; Almeida, F.D.L.; Cavalcante, R.S.; de Brito, E.S.; Cullen, P.J.; Frias, J.M.; Bourke, P.; Fernandes, F.A.N.; Rodrigues, S. H-1 NMR spectroscopy and chemometrics evaluation of non-thermal processing of orange juice. *Food Chem.* **2016**, *204*, 102–107. [CrossRef] [PubMed]

58. Spraul, M.; Link, M.; Schaefer, H.; Fang, F.; Schuetz, B. Wine analysis to check quality and authenticity by fully-automated H-1-NMR. In *38th World Congress of Vine and Wine*; JeanMarie, A., Ed.; EDP Sciences: Les Ulis, France, 2015; Volume 5.

59. Mulas, G.; Galaffu, M.G.; Pretti, L.; Nieddu, G.; Mercenaro, L.; Tonelli, R.; Anedda, R. NMR analysis of seven selections of vermentino grape berry: Metabolites composition and development. *J. Agric. Food Chem.* **2011**, *59*, 793–802. [CrossRef] [PubMed]

60. Picone, G.; Trimigno, A.; Tessarin, P.; Donnini, S.; Rombola, A.D.; Capozzi, F. H-1 nmr foodomics reveals that the biodynamic and the organic cultivation managements produce different grape berries (*Vitis vinifera* L. cv. Sangiovese). *Food Chem.* **2016**, *213*, 187–195. [CrossRef] [PubMed]

61. Mallamace, D.; Longo, S.; Corsaro, C. Proton NMR study of extra virgin olive oil with temperature: Freezing and melting kinetics. *Physica A* **2018**, *499*, 20–27. [CrossRef]

62. Ozdemir, I.S.; Dag, C.; Makuc, D.; Ertas, E.; Plavec, J.; Bekiroglu, S. Characterisation of the turkish and slovenian extra virgin olive oils by chemometric analysis of the presaturation h-1 nmr spectra. *LWT-Food Sci. Technol.* **2018**, *92*, 10–15. [CrossRef]

63. Mannina, L.; Marini, F.; Antiochia, R.; Cesa, S.; Magr, A.; Capitani, D.; Sobolev, A.P. Tracing the origin of beer samples by NMR and chemometrics: Trappist beers as a case study. *Electrophoresis* **2016**, *37*, 2710–2719. [CrossRef] [PubMed]

64. Sanchez-Estebanez, C.; Ferrero, S.; Alvarez, C.M.; Villafane, F.; Caballero, I.; Blanco, C.A. Nuclear magnetic resonance methodology for the analysis of regular and non-alcoholic lager beers. *Food Anal. Meth.* **2018**, *11*, 11–22. [CrossRef]

65. Bingol, K.; Brüschweiler, R. Two elephants in the room: New hybrid nuclear magnetic resonance and mass spectrometry approaches for metabolomics. *Curr. Opin. Clin. Nutr. Metab. Care* **2015**, *18*, 471–477. [CrossRef] [PubMed]

66. Xia, J.; Sinelnikov, I.V.; Han, B.; Wishart, D.S. Metaboanalyst 3.0-making metabolomics more meaningful. *Nucleic Acids Res.* **2015**, *43*, W251–W257. [CrossRef] [PubMed]

67. Forsberg, E.M.; Huan, T.; Rinehart, D.; Benton, H.P.; Warth, B.; Hilmers, B.; Siuzdak, G. Data processing, multi-omic pathway mapping, and metabolite activity analysis using XCMS online. *Nat. Protoc.* **2018**, *13*, 633–651. [CrossRef] [PubMed]

68. Spicer, R.; Salek, R.M.; Moreno, P.; Cañueto, D.; Steinbeck, C. Navigating freely-available software tools for metabolomics analysis. *Metabolomics* **2017**, *13*, 106. [CrossRef] [PubMed]

69. Ellinger, J.J.; Chylla, R.A.; Ulrich, E.L.; Markley, J.L. Databases and software for NMR-based metabolomics. *Curr. Metab.* **2013**, *1*. [CrossRef]

70. Lewis, I.A.; Shortreed, M.R.; Hegeman, A.D.; Markley, J.L. *Novel NMR and MS Approaches to Metabolomics*; Fan, T., Higashi, R.M., Lane, A.N., Eds.; Springer: New York, NY, USA, 2012; Volume 17, pp. 199–230.

71. Aggio, R.; Villas-Bôas, S.G.; Ruggiero, K. Metab: An R package for high-throughput analysis of metabolomics data generated by GC-MS. *Bioinformatics* **2011**, *27*, 2316–2318. [CrossRef] [PubMed]

72. Kirpich, A.S.; Ibarra, M.; Moskalenko, O.; Fear, J.M.; Gerken, J.; Mi, X.; Ashrafi, A.; Morse, A.M.; McIntyre, L.M. Secimtools: A suite of metabolomics data analysis tools. *BMC Bioinform.* **2018**, *19*, 151. [CrossRef] [PubMed]

73. Cambiaghi, A.; Ferrario, M.; Masseroli, M. Analysis of metabolomic data: Tools, current strategies and future challenges for omics data integration. *Brief. Bioinform.* **2017**, *18*, 498–510. [CrossRef] [PubMed]

74. Uppal, K.; Soltow, Q.A.; Promislow, D.E.L.; Wachtman, L.M.; Quyyumi, A.A.; Jones, D.P. Metabnet: An R package for metabolic association analysis of high-resolution metabolomics data. *Front. Bioeng. Biotechnol.* **2015**, *3*, 87. [CrossRef] [PubMed]

75. Uppal, K.; Walker, D.I.; Jones, D.P. Xmsannotator: An R package for network-based annotation of high-resolution metabolomics data. *Anal. Chem.* **2017**, *89*, 1063–1067. [CrossRef] [PubMed]

76. Aggio, R.B.M.; Ruggiero, K.; Villas-Bôas, S.G. Pathway activity profiling (PAPI): From the metabolite profile to the metabolic pathway activity. *Bioinformatics* **2010**, *26*, 2969–2976. [CrossRef] [PubMed]

77. Helmus, J.J.; Jaroniec, C.P. Nmrglue: An open source python package for the analysis of multidimensional nmr data. *J. Biomol. NMR* **2013**, *55*, 355–367. [CrossRef] [PubMed]

78. Pedregosa, F.; Weiss, R.; Brucher, M. Scikit-learn: Machine learning in python. *J. Mach. Learn. Res.* **2011**, *12*, 2825–2830.

79. Warth, B.; Levin, N.; Rinehart, D.; Teijaro, J.; Benton, H.P.; Siuzdak, G. Metabolizing data in the cloud. *Trends Biotechnol.* **2017**, *35*, 481–483. [CrossRef] [PubMed]

80. Chong, J.; Soufan, O.; Li, C.; Caraus, I.; Li, S.; Bourque, G.; Wishart, D.S.; Xia, J. Metaboanalyst 4.0: Towards more transparent and integrative metabolomics analysis. *Nucleic Acids Res.* **2018**, *46*, W486–W494. [CrossRef] [PubMed]

81. Haug, K.; Salek, R.M.; Steinbeck, C. Global open data management in metabolomics. *Curr. Opin. Chem. Biol.* **2017**, *36*, 58–63. [CrossRef] [PubMed]

82. Haug, K.; Salek, R.M.; Conesa, P.; Hastings, J.; de Matos, P.; Rijnbeek, M.; Mahendraker, T.; Williams, M.; Neumann, S.; Rocca-Serra, P.; et al. Metabolights—An open-access general-purpose repository for metabolomics studies and associated meta-data. *Nucleic Acids Res* **2013**, *41*, D781–D786. [CrossRef] [PubMed]

83. Sud, M.; Fahy, E.; Cotter, D.; Azam, K.; Vadivelu, I.; Burant, C.; Edison, A.; Fiehn, O.; Higashi, R.; Nair, K.S.; et al. Metabolomics workbench: An international repository for metabolomics data and metadata, metabolite standards, protocols, tutorials and training, and analysis tools. *Nucleic Acids Res.* **2016**, *44*, D463–D470. [CrossRef] [PubMed]

84. Fiehn, O.; Robertson, D.; Griffin, J.; van der Werf, M.; Nikolau, B.; Morrison, N.; Sumner, L.W.; Goodacre, R.; Hardy, N.W.; Taylor, C.; et al. The metabolomics standards initiative (MSI). *Metabolomics* **2007**, *3*, 175–178. [CrossRef]

85. Salek, R.M.; Neumann, S.; Schober, D.; Hummel, J.; Billiau, K.; Kopka, J.; Correa, E.; Reijmers, T.; Rosato, A.; Tenori, L.; et al. Coordination of standards in metabolomics (cosmos): Facilitating integrated metabolomics data access. *Metabolomics* **2015**, *11*, 1587–1597. [CrossRef] [PubMed]

86. Trufelli, H.; Palma, P.; Famiglini, G.; Cappiello, A. An overview of matrix effects in liquid chromatography-mass spectrometry. *Mass Spectrom. Rev.* **2011**, *30*, 491–509. [CrossRef] [PubMed]

87. Cappiello, A.; Famiglini, G.; Palma, P.; Trufelli, H. Matrix effects in liquid chromatography-mass spectrometry. *J. Liquid Chromatogr. Relat. Technol.* **2010**, *33*, 1067–1081. [CrossRef]

88. Weber, C.M.; Cauchi, M.; Patel, M.; Bessant, C.; Turner, C.; Britton, L.E.; Willis, C.M. Evaluation of a gas sensor array and pattern recognition for the identification of bladder cancer from urine headspace. *Analyst* **2011**, *136*, 359–364. [CrossRef] [PubMed]

89. Klapková, E.; Uřinovská, R.; Průša, R. The influence of matrix effects on high performance liquid chromatography-mass spectrometry methods development and validation. *Vliv Matricových Efektů Vývoji* **2011**, *19*, 5–8.

90. Ye, Z.; Tsao, H.; Gao, H.; Brummel, C.L. Minimizing matrix effects while preserving throughput in LC-MS/MS bioanalysis. *Bioanalysis* **2011**, *3*, 1587–1601. [CrossRef] [PubMed]

91. Wang, L.Q.; Zeng, Z.L.; Su, Y.J.; Zhang, G.K.; Zhong, X.L.; Liang, Z.P.; He, L.M. Matrix effects in analysis of β-agonists with LC-MS/MS: Influence of analyte concentration, sample source, and SPE type. *J. Agric. Food Chem.* **2012**, *60*, 6359–6363. [CrossRef] [PubMed]

92. Peters, F.T.; Remane, D. Aspects of matrix effects in applications of liquid chromatography-mass spectrometry to forensic and clinical toxicology—A review. *Anal. Bioanal. Chem.* **2012**, *403*, 2155–2172. [CrossRef] [PubMed]

93. Jiang, H.; Cao, H.; Zhang, Y.; Fast, D.M. Systematic evaluation of supported liquid extraction in reducing matrix effect and improving extraction efficiency in LC-MS/MS based bioanalysis for 10 model pharmaceutical compounds. *J. Chromatogr. B* **2012**, *891–892*, 71–80. [CrossRef] [PubMed]

94. Villas-Bôas, S.G.; Koulman, A.; Lane, G.A. *Analytical Methods from the Perspective of Method Standardization*; Nielsen, J., Jewett, M.C., Eds.; Springer: Berlin/Heidelberg, Germany, 2007; Volume 18, pp. 11–52.

95. Ali, K.; Maltese, F.; Toepfer, R.; Choi, Y.H.; Verpoorte, R. Metabolic characterization of palatinate German white wines according to sensory attributes, varieties, and vintages using NMR spectroscopy and multivariate data analyses. *J. Biomol. NMR* **2011**, *49*, 255–266. [CrossRef] [PubMed]

96. Lee, J.E.; Hwang, G.S.; Van Den Berg, F.; Lee, C.H.; Hong, Y.S. Evidence of vintage effects on grape wines using 1H NMR-based metabolomic study. *Anal. Chim. Acta* **2009**, *648*, 71–76. [CrossRef] [PubMed]

97. Atanassov, I.; Hvarleva, T.; Rusanov, K.; Tsvetkov, I.; Atanassov, A. Wine metabolite profiling: Possible application in winemaking and grapevine breeding in bulgaria. *Biotechnol. Biotechnol. Equip.* **2009**, *23*, 1449–1452. [CrossRef]

98. Howell, K.S.; Cozzolino, D.; Bartowsky, E.J.; Fleet, G.H.; Henschke, P.A. Metabolic profiling as a tool for revealing *Saccharomyces* interactions during wine fermentation. *FEMS Yeast Res.* **2006**, *6*, 91–101. [CrossRef] [PubMed]

99. Tumanov, S.; Zubenko, Y.; Greven, M.; Greenwood, D.R.; Shmanai, V.; Villas-Boas, S.G. Comprehensive lipidome profiling of Sauvignon blanc grape juice. *Food Chem.* **2015**, *180*, 249–256. [CrossRef] [PubMed]

100. Pinto, J.; Oliveira, A.S.; Azevedo, J.; De Freitas, V.; Lopes, P.; Roseira, I.; Cabral, M.; de Pinho, P.G. Assessment of oxidation compounds in oaked chardonnay wines: A GC-MS and H-1 NMR metabolomics approach. *Food Chem.* **2018**, *257*, 120–127. [CrossRef] [PubMed]

101. Arita, K.; Honma, T.; Suzuki, S. Comprehensive and comparative lipidome analysis of vitis vinifera l. Cv. Pinot noir and japanese indigenous v. Vinifera l. Cv. Koshu grape berries. *PLoS ONE* **2017**, *12*, e0186952. [CrossRef] [PubMed]

102. Tejero Rioseras, A.; Garcia Gomez, D.; Ebert, B.E.; Blank, L.M.; Ibáñez, A.J.; Sinues, P.M.L. Comprehensive real-time analysis of the yeast volatilome. *Sci. Rep.* **2017**, *7*, 14236. [CrossRef] [PubMed]

103. Amann, A.; Costello, B.L.; Miekisch, W.; Schubert, J.; Buszewski, B.; Pleil, J.; Ratcliffe, N.; Risby, T. The human volatilome: Volatile organic compounds (VOCs) in exhaled breath, skin emanations, urine, feces and saliva. *J. Breath Res.* **2014**, *8*, 034001. [CrossRef] [PubMed]

104. Opitz, P.; Herbarth, O. The volatilome—Investigation of volatile organic metabolites (vom) as potential tumor markers in patients with head and neck squamous cell carcinoma (hnscc). *J. Otolaryngol. Head Neck Surg.* **2018**, *47*, 42. [CrossRef] [PubMed]

105. Skogerson, K.; Runnebaum, R.O.N.; Wohlgemuth, G.; De Ropp, J.; Heymann, H.; Fiehn, O. Comparison of gas chromatography-coupled time-of-flight mass spectrometry and 1H nuclear magnetic resonance spectroscopy metabolite identification in white wines from a sensory study investigating wine body. *J. Agric. Food Chem.* **2009**, *57*, 6899–6907. [CrossRef] [PubMed]

106. Zott, K.; Thibon, C.; Bely, M.; Lonvaud-Funel, A.; Dubourdieu, D.; Masneuf-Pomarede, I. The grape must non-*Saccharomyces* microbial community: Impact on volatile thiol release. *Int. J. Food Microbiol.* **2011**, *151*, 210–215. [CrossRef] [PubMed]

107. Pinu, F.R.; Edwards, P.J.B.; Gardner, R.C.; Villas-Boas, S.G. Nitrogen and carbon assimilation by *Saccharomyces cerevisiae* during Sauvignon blanc juice fermentation. *FEMS Yeast Res.* **2014**, *14*, 1206–1222. [CrossRef] [PubMed]

108. Casu, F.; Pinu, F.R.; Fedrizzi, B.; Greenwood, D.R.; Villas-Boas, S.G. The effect of linoleic acid on the Sauvignon blanc fermentation by different wine yeast strains. *FEMS Yeast Res.* **2016**, *16*, fow050. [CrossRef] [PubMed]

109. Schueuermann, C.; Steel, C.C.; Blackman, J.W.; Clark, A.C.; Schwarz, L.J.; Moraga, J.; Collado, I.G.; Schmidtke, L.M. A GC-MS untargeted metabolomics approach for the classification of chemical differences in grape juices based on fungal pathogen. *Food Chem.* **2019**, *270*, 375–384. [CrossRef] [PubMed]

110. Cuadros-Inostroza, A.; Ruíz-Lara, S.; González, E.; Eckardt, A.; Willmitzer, L.; Peña-Cortés, H. Gc–ms metabolic profiling of cabernet sauvignon and merlot cultivars during grapevine berry development and network analysis reveals a stage- and cultivar-dependent connectivity of primary metabolites. *Metabolomics* **2016**, *12*, 39. [CrossRef] [PubMed]

111. Theodoridis, G.; Gika, H.; Franceschi, P.; Caputi, L.; Arapitsas, P.; Scholz, M.; Masuero, D.; Wehrens, R.; Vrhovsek, U.; Mattivi, F. LC-MS based global metabolite profiling of grapes: Solvent extraction protocol optimisation. *Metabolomics* **2012**, *8*, 175–185. [CrossRef]

112. Capone, D.L.; Pardon, K.H.; Cordente, A.G.; Jeffery, D.W. Identification and quantitation of 3-s-cysteinylglycinehexan-1-ol (cysgly-3-mh) in sauvignon blanc grape juice by HPLC-MS/MS. *J. Agric. Food Chem.* **2011**, *59*, 11204–11210. [CrossRef] [PubMed]

113. Capone, D.L.; Black, C.A.; Jeffery, D.W. Effects on 3-mercaptohexan-1-ol precursor concentrations from prolonged storage of Sauvignon blanc grapes prior to crushing and pressing. *J. Agric. Food Chem.* **2012**, *60*, 3515–3523. [CrossRef] [PubMed]

114. Arapitsas, P.; Della Corte, A.; Gika, H.; Narduzzi, L.; Mattivi, F.; Theodoridis, G. Studying the effect of storage conditions on the metabolite content of red wine using HILIC LC-MS based metabolomics. *Food Chem.* **2016**, *197*, 1331–1340. [CrossRef] [PubMed]

115. Diaz, R.; Gallart-Ayala, H.; Sancho, J.V.; Nunez, O.; Zamora, T.; Martins, C.P.B.; Hernandez, F.; Hernandez-Cassou, S.; Saurina, J.; Checa, A. Told through the wine: A liquid chromatography-mass spectrometry interplatform comparison reveals the influence of the global approach on the final annotated metabolites in non-targeted metabolomics. *J. Chromatogr. A* **2016**, *1433*, 90–97. [CrossRef] [PubMed]

116. Roullier-Gall, C.; Hemmler, D.; Witting, M.; Moritz, F.; Heinzmann, S.; Jeandet, P.; Gonsior, M.; Gougeon, R.; Schmitt-Kopplin, P. Metabolomics characterization of bottled wine: Impact of environmental parameters. In *Abstracts of Papers of the American Chemical Society*; American Chemical Society: Washington, DC, USA, 2016; Volume 252.

117. Roullier-Gall, C.; Witting, M.; Gougeon, R.D.; Schmitt-Kopplin, P. High precision mass measurements for wine metabolomics. *Front. Chem.* **2014**, *2*, 102. [CrossRef] [PubMed]

118. Roullier-Gall, C.; Witting, M.; Tziotis, D.; Ruf, A.; Gougeon, R.D.; Schmitt-Kopplin, P. Integrating analytical resolutions in non-targeted wine metabolomics. *Tetrahedron* **2015**, *71*, 2983–2990. [CrossRef]

119. Ruocco, S.; Stefanini, M.; Stanstrup, J.; Perenzoni, D.; Mattivi, F.; Vrhovsek, U. The metabolomic profile of red non-v-vinifera genotypes. *Food Res. Int.* **2017**, *98*, 10–19. [CrossRef] [PubMed]

120. Son, H.S.; Hwang, G.S.; Kim, K.M.; Ahn, H.J.; Park, W.M.; Van Den Berg, F.; Hong, Y.S.; Lee, C.H. Metabolomic studies on geographical grapes and their wines using 1H NMR analysis coupled with multivariate statistics. *J. Agric. Food Chem.* **2009**, *57*, 1481–1490. [CrossRef] [PubMed]

121. Rochfort, S.; Ezernieks, V.; Bastian, S.E.P.; Downey, M.O. Sensory attributes of wine influenced by variety and berry shading discriminated by NMR metabolomics. *Food Chem.* **2010**, *121*, 1296–1304. [CrossRef]

122. Hong, Y.S.; Cilindre, C.; Liger-Belair, G.; Jeandet, P.; Hertkorn, N.; Schmitt-Kopplin, P. Metabolic influence of *Botrytis cinerea* infection in champagne base wine. *J. Agric. Food Chem.* **2011**, *59*, 7237–7245. [CrossRef] [PubMed]

123. Zhu, J.Y.; Hu, B.R.; Lu, J.; Xu, S.C. Analysis of metabolites in Cabernet sauvignon and shiraz dry red wines from Shanxi by H1 NMR spectroscopy combined with pattern recognition analysis. *Open Chem.* **2018**, *16*, 446–452. [CrossRef]

124. Simó, C.; Moreno-Arribas, M.V.; Cifuentes, A. Ion-trap versus time-of-flight mass spectrometry coupled to capillary electrophoresis to analyze biogenic amines in wine. *J. Chromatogr. A* **2008**, *1195*, 150–156. [CrossRef] [PubMed]

125. Vanhoenacker, G.; De Villiers, A.; Lazou, K.; De Keukeleire, D.; Sandra, P. Comparison of high-performance liquid chromatography-mass spectroscopy and capillary electrophoresis-mass spectroscopy for the analysis of phenolic compounds in diethyl ether extracts of red wines. *Chromatographia* **2001**, *54*, 309–315. [CrossRef]

126. Wine-Omics. Available online: https://www.nature.com/articles/455699a (accessed on 2 October 2008).

127. Arapitsas, P.; Scholz, M.; Vrhovsek, U.; Di Blasi, S.; Biondi Bartolini, A.; Masuero, D.; Perenzoni, D.; Rigo, A.; Mattivi, F. A metabolomic approach to the study of wine micro-oxygenation. *PLoS ONE* **2012**, *7*, e37783. [CrossRef] [PubMed]

128. Liu, Y.Z.; Forcisi, S.; Harir, M.; Deleris-Bou, M.; Krieger-Weber, S.; Lucio, M.; Longin, C.; Degueurce, C.; Gougeon, R.D.; Schmitt-Kopplin, P.; et al. New molecular evidence of wine yeast-bacteria interaction unraveled by non-targeted exometabolomic profiling. *Metabolomics* **2016**, *12*, 16. [CrossRef]

129. Conterno, L.; Aprea, E.; Franceschi, P.; Viola, R.; Vrhovsek, U. Overview of *Dekkera bruxellensis* behaviour in an ethanol-rich environment using untargeted and targeted metabolomic approaches. *Food Res. Int.* **2013**, *51*, 670–678. [CrossRef]

130. Rocchetti, G.; Gatti, M.; Bavaresco, L.; Lucini, L. Untargeted metabolomics to investigate the phenolic composition of Chardonnay wines from different origins. *J. Food Compos. Anal.* **2018**, *71*, 87–93. [CrossRef]

131. López-Rituerto, E.; Savorani, F.; Avenoza, A.; Busto, J.H.; Peregrina, J.M.; Engelsen, S.B. Investigations of la rioja terroir for wine production using 1H NMR metabolomics. *J. Agric. Food Chem.* **2012**, *60*, 3452–3461. [CrossRef] [PubMed]

132. Amargianitaki, M.; Spyros, A. NMR-based metabolomics in wine quality control and authentication. *Chem. Biol. Technol. Agric.* **2017**, *4*, 9. [CrossRef]
133. Cassino, C.; Tsolakis, C.; Bonello, F.; Gianotti, V.; Osella, D. Effects of area, year and climatic factors on barbera wine characteristics studied by the combination of 1H-NMR metabolomics and chemometrics. *J. Wine Res.* **2017**, *28*, 259–277. [CrossRef]
134. Roullier-Gall, C.; Boutegrabet, L.; Gougeon, R.D.; Schmitt-Kopplin, P. A grape and wine chemodiversity comparison of different appellations in burgundy: Vintage vs *terroir* effects. *Food Chem.* **2014**, *152*, 100–107. [CrossRef] [PubMed]
135. Roullier-Gall, C.; Lucio, M.; Noret, L.; Schmitt-Kopplin, P.; Gougeon, R.D. How subtle is the "*terroir*" effect? Chemistry-related signatures of two "climats de bourgogne". *PLoS ONE* **2014**, *9*, e97615. [CrossRef] [PubMed]
136. Bokulich, N.A.; Collins, T.S.; Masarweh, C.; Allen, G.; Heymann, H.; Ebeler, S.E.; Mills, D.A. Associations among wine grape microbiome, metabolome, and fermentation behavior suggest microbial contribution to regional wine characteristics. *mBio* **2016**, *7*. [CrossRef] [PubMed]
137. Anesi, A.; Stocchero, M.; Dal Santo, S.; Commisso, M.; Zenoni, S.; Ceoldo, S.; Tornielli, G.B.; Siebert, T.E.; Herderich, M.; Pezzotti, M.; et al. Towards a scientific interpretation of the terroir concept: Plasticity of the grape berry metabolome. *BMC Plant Biol.* **2015**, *15*, 191. [CrossRef] [PubMed]
138. King, E.S.; Kievit, R.L.; Curtin, C.; Swiegers, J.H.; Pretorius, I.S.; Bastian, S.E.P.; Leigh Francis, I. The effect of multiple yeasts co-inoculations on Sauvignon blanc wine aroma composition, sensory properties and consumer preference. *Food Chem.* **2010**, *122*, 618–626. [CrossRef]
139. Ciani, M.; Capece, A.; Comitini, F.; Canonico, L.; Siesto, G.; Romano, P. Yeast interactions in inoculated wine fermentation. *Front. Microbiol.* **2016**, *7*, 555. [CrossRef] [PubMed]
140. Pizarro, F.; Vargas, F.A.; Agosin, E. A systems biology perspective of wine fermentations. *Yeast* **2007**, *24*, 977–991. [CrossRef] [PubMed]
141. Tumanov, S.; Pinu, F.R.; Greenwood, D.R.; Villas-Boas, S.G. The effect of free fatty acids and lipolysis on sauvignon blanc fermentation. *Aust. J. Grape Wine Res.* **2017**, in press. [CrossRef]
142. Aliverdieva, D.A.; Mamaev, D.V.; Bondarenko, D.I.; Sholtz, K.F. Properties of yeast *Saccharomyces cerevisiae* plasma membrane dicarboxylate transporter. *Biochemistry* **2006**, *71*, 1161–1169. [CrossRef] [PubMed]
143. Crépin, L.; Nidelet, T.; Sanchez, I.; Dequin, S.; Camarasa, C. Sequential use of nitrogen compounds by yeast during wine fermentation: A model based on kinetic and regulation characteristics of nitrogen permeases. *Appl. Environ. Microbiol.* **2012**, in press.
144. Cooper, T.G. Nitrogen metabolism in *Saccharomyces cerevisiae*. In *The Molecular Biology of the Yeast Saccharomyces: Metabolism and Gene Expression*; Strathern, J.N., Jones, E.W., Broach, J.R., Eds.; Cold Spring Harbor Laboratory Press: Cold Spring Harbor, NY, USA, 1982; pp. 39–99.
145. Hofman-Bang, J. Nitrogen catabolite repression in saccharomyces cerevisiae. *Mol. Biotechnol.* **1999**, *12*, 35–73. [CrossRef]
146. Henscke, P.; Jiranek, V. Yeasts—Metabolism of nitrogen compounds. In *Wine Microbiology, Biotechnology*; Harwood Academic: Chur, Switzerland, 1993; pp. 77–164.
147. Fairbairn, S.; McKinnon, A.; Musarurwa, H.T.; Ferreira, A.C.; Bauer, F.F. The impact of single amino acids on growth and volatile aroma production by *Saccharomyces cerevisiae* strains. *Front. Microbiol.* **2017**, *8*, 2554. [CrossRef] [PubMed]
148. Peltier, E.; Bernard, M.; Trujillo, M.; Prodhomme, D.; Barbe, J.-C.; Gibon, Y.; Marullo, P. Wine yeast phenomics: A standardized fermentation method for assessing quantitative traits of *Saccharomyces cerevisiae* strains in enological conditions. *PLoS ONE* **2018**, *13*, e0190094. [CrossRef] [PubMed]
149. Varela, C.; Torrea, D.; Schmidt, S.A.; Ancin-Azpilicueta, C.; Henschke, P.A. Effect of oxygen and lipid supplementation on the volatile composition of chemically defined medium and chardonnay wine fermented with *Saccharomyces cerevisiae*. *Food Chem.* **2012**, *135*, 2863–2871. [CrossRef] [PubMed]
150. López-Malo, M.; Querol, A.; Guillamon, J.M. Metabolomic comparison of *Saccharomyces cerevisiae* and the cryotolerant species *S. bayanus var. uvarum* and *S. kudriavzevii* during wine fermentation at low temperature. *PLoS ONE* **2013**, *8*, e60135. [CrossRef] [PubMed]
151. Pizarro, F.J.; Jewett, M.C.; Nielsen, J.; Agosin, E. Growth temperature exerts differential physiological and transcriptional responses in laboratory and wine strains of *Saccharomyces cerevisiae*. *Appl. Environ. Microbiol.* **2008**, *74*, 6358. [CrossRef] [PubMed]

152. Rollero, S.; Bloem, A.; Camarasa, C.; Sanchez, I.; Ortiz-Julien, A.; Sablayrolles, J.-M.; Dequin, S.; Mouret, J.-R. Combined effects of nutrients and temperature on the production of fermentative aromas by *Saccharomyces cerevisiae* during wine fermentation. *Appl. Microbiol. Biotechnol.* **2015**, *99*, 2291–2304. [CrossRef] [PubMed]

153. Aceituno, F.F.; Orellana, M.; Torres, J.; Mendoza, S.; Slater, A.W.; Melo, F.; Agosin, E. Oxygen response of the wine yeast *Saccharomyces cerevisiae* EC1118 grown under carbon-sufficient, nitrogen-limited enological conditions. *Appl. Environ. Microbiol.* **2012**, *78*, 8340. [CrossRef] [PubMed]

154. Orellana, M.; Aceituno, F.F.; Slater, A.W.; Almonacid, L.I.; Melo, F.; Agosin, E. Metabolic and transcriptomic response of the wine yeast *Saccharomyces cerevisiae* strain ec1118 after an oxygen impulse under carbon-sufficient, nitrogen-limited fermentative conditions. *FEMS Yeast Res.* **2014**, *14*, 412–424. [CrossRef] [PubMed]

155. Mendes, I.; Sanchez, I.; Franco-Duarte, R.; Camarasa, C.; Schuller, D.; Dequin, S.; Sousa, M.J. Integrating transcriptomics and metabolomics for the analysis of the aroma profiles of *Saccharomyces cerevisiae* strains from diverse origins. *BMC Genomics* **2017**, *18*, 13. [CrossRef] [PubMed]

156. Hochberg, U.; Degu, A.; Cramer, G.R.; Rachmilevitch, S.; Fait, A. Cultivar specific metabolic changes in grapevines berry skins in relation to deficit irrigation and hydraulic behavior. *Plant Physiol. Biochem.* **2015**, *88*, 42–52. [CrossRef] [PubMed]

157. Pinasseau, L.; Vallverdu-Queralt, A.; Verbaere, A.; Roques, M.; Meudec, E.; Le Cunff, L.; Peros, J.P.; Ageorges, A.; Sommerer, N.; Boulet, J.C.; et al. Cultivar diversity of grape skin polyphenol composition and changes in response to drought investigated by LC-MS based metabolomics. *Front. Plant Sci.* **2017**, *8*, 24. [CrossRef] [PubMed]

158. Brilli, M.; Asquini, E.; Moser, M.; Bianchedi, P.L.; Perazzolli, M.; Si-Ammour, A. A multi-omics study of the grapevine-downy mildew (*Plasmopara viticola*) pathosystem unveils a complex protein coding- and noncoding-based arms race during infection. *Sci. Rep.* **2018**, *8*, 757. [CrossRef] [PubMed]

159. Benheim, D.; Rochfort, S.; Ezernieks, V.; Korosi, G.A.; Powell, K.S.; Robertson, E.; Potter, I.D. Early detection of grape phylloxera (daktulosphaira vitifoliae fitch) infestation through identification of chemical biomarkers. In *V international Phylloxera Symposium*; Griesser, M., Forneck, A., Eds.; International Society for Horticultural Science: Leuven, Belgium, 2011; Volume 904, pp. 17–24.

160. Benheim, D.; Rochfort, S.; Korosi, G.A.; Powell, K.S.; Robertson, E.; Potter, I.D. Nuclear magnetic resonance metabolic profiling of leaves from vitis vinifera infested with root-feeding grape phylloxera (daktulosphaira vitifoliae fitch) under field conditions. In *VI International Phylloxera Symposium*; Ollat, N., Papura, D., Eds.; International Society for Horticultural Science: Leuven, Belgium, 2014; Volume 1045, pp. 59–66.

161. Wishart, D.S. Advances in metabolite identification. *Bioanalysis* **2011**, *3*, 1769–1782. [CrossRef] [PubMed]

162. Date, Y.; Kikuchi, J. Application of a deep neural network to metabolomics studies and its performance in determining important variables. *Anal. Chem.* **2018**, *90*, 1805–1810. [CrossRef] [PubMed]

163. Trivedi, D.K.; Hollywood, K.A.; Goodacre, R. Metabolomics for the masses: The future of metabolomics in a personalized world. *New Horiz. Transl. Med.* **2017**, *3*, 294–305. [CrossRef] [PubMed]

164. Scalbert, A.; Brennan, L.; Manach, C.; Andres-Lacueva, C.; Dragsted, L.O.; Draper, J.; Rappaport, S.M.; Van Der Hooft, J.J.J.; Wishart, D.S. The food metabolome: A window over dietary exposure. *Am. J. Clin. Nutr.* **2014**, *99*, 1286–1308. [CrossRef] [PubMed]

165. Förster, J.; Famili, I.; Fu, P.; Palsson, B.Ø.; Nielsen, J. Genome-scale reconstruction of the saccharomyces cerevisiae metabolic network. *Genome Res.* **2003**, *13*, 244–253. [CrossRef] [PubMed]

166. Zhang, C.; Hua, Q. Applications of genome-scale metabolic models in biotechnology and systems medicine. *Front. Physiol.* **2016**, *6*, 413. [CrossRef] [PubMed]

167. Kerkhoven, E.J.; Lahtvee, P.-J.; Nielsen, J. Applications of computational modeling in metabolic engineering of yeast. *FEMS Yeast Res.* **2015**, *15*, 1–13. [CrossRef] [PubMed]

168. Heavner, B.D.; Price, N.D. Comparative analysis of yeast metabolic network models highlights progress, opportunities for metabolic reconstruction. *PLOS Comput. Biol.* **2015**, *11*, e1004530. [CrossRef] [PubMed]

169. Mendoza, S.N.; Cañón, P.M.; Contreras, Á.; Ribbeck, M.; Agosín, E. Genome-scale reconstruction of the metabolic network in *Oenococcus oeni* to assess wine malolactic fermentation. *Front. Microbiol.* **2017**, *8*, 534. [CrossRef] [PubMed]

170. Saitua, F.; Torres, P.; Pérez-Correa, J.R.; Agosin, E. Dynamic genome-scale metabolic modeling of the yeast *Pichia pastoris*. *BMC Syst. Biol.* **2017**, *11*, 27. [CrossRef] [PubMed]

171. Pizarro, F.; Varela, C.; Martabit, C.; Bruno, C.; Perez-Correa, J.R.; Agosin, E. Coupling kinetic expressions and metabolic networks for predicting wine fermentations. *Biotechnol. Bioeng.* **2007**, *98*, 986–998. [CrossRef] [PubMed]

172. Vargas, F.A.; Pizarro, F.; Pérez-Correa, J.R.; Agosin, E. Expanding a dynamic flux balance model of yeast fermentation to genome-scale. *BMC Syst. Biol.* **2011**, *5*, 75. [CrossRef] [PubMed]

fermentation

MDPI

Review

Recent Developments on the Origin and Nature of Reductive Sulfurous Off-Odours in Wine

Nikolaus Müller [1,*] and Doris Rauhut [2]

[1]   Silvanerweg 9, 55595 Wallhausen, Germany
[2]   Department of Microbiology and Biochemistry, Hochschule Geisenheim University, Von-Lade-Straße 1,
       65366 Geisenheim, Germany; doris.rauhut@hs-gm.de
*   Correspondence: nik.mueller@t-online.de; Tel.: +49-6706-913103

Received: 9 July 2018; Accepted: 3 August 2018; Published: 8 August 2018

Abstract: Reductive sulfurous off-odors are still one of the main reasons for rejecting wines by consumers. In 2008 at the International Wine Challenge in London, approximately 6% of the more than 10,000 wines presented were described as faulty. Twenty-eight percent were described as faulty because they presented "reduced characters" similar to those presented by "cork taint" and in nearly the same portion. Reductive off-odors are caused by low volatile sulfurous compounds. Their origin may be traced back to the metabolism of the microorganisms (yeasts and lactic acid bacteria) involved in the fermentation steps during wine making, often followed by chemical conversions. The main source of volatile sulfur compounds (VSCs) are precursors from the sulfate assimilation pathway (SAP, sometimes named as the "sulfate reduction pathway" SRP), used by yeast to assimilate sulfur from the environment and incorporate it into the essential sulfur-containing amino acids methionine and cysteine. Reductive off-odors became of increasing interest within the last few years, and the method to remove them by treatment with copper (II) salts (sulfate or citrate) is more and more questioned: The effectiveness is doubted, and after prolonged bottle storage, they reappear quite often. Numerous reports within the last few years and an ongoing flood of publications dealing with this matter reflect the importance of this problem. In a recent detailed review, almost all relevant aspects were discussed on a scientific data basis, and a "decision tree" was formulated to support winemakers handling this problem. Since we are dealing with a very complicated matter with a multitude of black spots still remaining, these advices can only be realized using specific equipment and special chemicals, not necessarily found in small wineries. The main problem in dealing with sulfurous compounds arises from the high variability of their reactivities. Sulfur is a metalloid with a large valence span across eight electron transformations from S (−II) up to S (+VI). This allows it to participate in an array of oxidation, reduction and disproportionation reactions, both abiotic and linked to microbial metabolism. In addition, sulfur is the element with the most allotropes and a high tendency to form chains and rings, with different stabilities of defined species and a high interconvertibility among each other. We suppose, there is simply a lack of knowledge of what is transferred during filling into bottles after fermentation and fining procedures. The treatment with copper (II) salts to remove sulfurous off-odors before filling rather increases instead of solving the problem. This paper picks up the abundant knowledge from recent literature and tries to add some aspects and observations, based on the assumption that the formation of polythionates, hitherto not taken into consideration, may explain some of the mystery of the re-appearance of reductive off-odors.

Keywords: reductive off-odors; reappearance; wine; volatile sulfur compounds; polythionates as precursors; elemental sulfur

1. Introduction

Reductive off-odor problems are mostly caused by volatile sulfur compounds (VSCs) and equal to cork taint are responsible for an important proportion (30%) of faulty wines [1]. H_2S is the most frequently found of these compounds, followed by methanethiol (MeSH) and ethanethiol (EtSH) [2]. Hydrogen sulfide can be generated by *Saccharomyces cerevisiae* in the course of sulfur assimilation during fermentation. Inorganic sources used in this "sulfate assimilation pathway" (SAP) are mainly sulfate, naturally occurring in the grape must, sulfite, usually added as an antioxidant and microbial agent [3] to control microbial spoilage of must and wine, and elemental sulfur, which is used as a fungicide in the vineyard [4–6]. To date, yeast genetic mechanisms of H_2S liberation during wine fermentation are well understood, and yeast strains producing low levels of H_2S have been developed [7,8]. As outlined later on in the case of low nitrogen status in must, there is a lack of the activated precursors of the sulfur-containing amino acids cysteine and methionine (O-acetylserine and -homoserine). The surplus of sulfide then is excreted as H_2S, which can react biochemically or chemically with other VSCs.

Other sources of VSCs are the enzymatic or chemical degradation of cysteine and methionine. Chemical degradation takes place via a Strecker reaction [9] and partly via Ehrlich degradation [10]. Strecker degradation leads to products of reactions between α-dicarbonyl compounds like diacetyl or o-quinones, easily formed from wine polyphenols by oxidation, and amino acids present in wine. In the case of methionine, methional is formed, which itself is considered as an aroma component of wine, but also easily cleaves off MeSH. During Ehrlich degradation, amino acids are transformed by a transamination reaction, followed by decarboxylation to methional, as well.

On the other hand, H_2S does not only serve as a reaction partner in the formation of sulfur-containing amino acids.

More and more attention is given to other functions of H_2S. It is now not merely recognized as intermediate in the biosynthesis of the sulfur-containing amino acids, but it has important functions as a potent biological effector, i.e., in detoxification, population, signaling and extending the life-span of microorganisms and mammalian cells [7,11,12].

Winemakers commonly add copper (II) salts to wine before bottling to remove these unpleasant reductive aromas. Those allowed are copper sulfate [13] or copper citrate [14]. Due to increasing problems due to the reappearance of sulfurous (reductive) off-flavors during storage, OIV (Organisation Internationale de la Vigne et du Vin) recently allowed the use of silver chloride as an alternative [15], although little is known about the side reactions of this treatment.

Indeed, the reappearance of reductive off-odors has been one of the main topics in recent wine research. Numerous literature reports on the increases of free H_2S and MeSH during wine storage have been published. A survey is given in [2].

In many of these articles, the addition of copper (II) salt to remove reduced aromas is made responsible for this phenomenon, and some striking theories have been developed. Nevertheless the factors for the appearance of reductive off-flavors in the bottle still remain arguably the most mysterious [16]. The main problem seems to be the lack of knowledge of what is transferred after a long wine processing storage in the bottles. In addition, it has to be kept in mind that the two chemical elements dealt with in the context of the reappearance of reductive off-odors belong simultaneously to those with a very "difficult" chemistry: the transition metal copper and the non-metal element sulfur. Therefore, some of the mystery might be elucidated by going back to the basic chemistry of these elements.

2. Origin of Reductive Off-Odors

In this article, according to the recent review of Kreitman et al. [2], the latent precursors and pathways likely to be responsible for the loss and formation of these sulfhydryls during wine storage based on the existing enology literature, as well as studies from food chemistry, geochemistry, biochemistry and synthetic chemistry have been evaluated and their findings have been expanded to substance classes, hitherto not taken into consideration.

In addition to these recently-proposed precursor classes [2], which have a sufficient concentration and metastability to serve as latent sulfhydryl precursors in wine, other VSCs forming precursor classes are considered:

(1) Elemental sulfur [4,5] has been used as a pesticide since antiquity and is still used for the purpose of fighting fungal diseases in grapes. Elemental sulfur in all its allotropes and modifications is not by far inert and participates considerably in the formation of precursors like polysulfanes and polythionates.

(2) Copper-sulfhydryl complexes formed by the addition of copper (II) salts. These release through an unknown mechanism of VSCs under reductive conditions [2,17].

(3) Disulfides (symmetrical and asymmetrical), polysulfur compounds with chains and rings as listed in Table 1 (polysulfanes and (di)organopolysulfanes), cleaved by thiolysis or sulfitolysis reactions [2].

(4) From polythionates formed by the reaction of yeast-borne hydrogen sulfide with added sulfur dioxide ('Wackenrodersche Flüssigkeit", from here on called "Wackenroder solution" (WS)).

(5) By hydrolysis of *S*-alkylthioacetates, by-products of alcoholic fermentation [2,18,19].

(6) Some evidence also exists for S-amino acids serving as precursors, as described below:

(a) Enzymatically by lyase activity of C-S bond-cleaving enzymes, dependent on the cofactor pyridoxal phosphate [20].

(b) Chemically by Strecker degradation of sulfur-containing amino acids [9].

(c) Pyrophosphate-catalyzed decarboxylation tartly enzymatically starting with Ehrlich degradation (transamination, followed by followed by thiaminpyrophosphate catalyzed decarboxylation to aldehydes). The latter may decompose by chemical cleavage of an unstable intermediate [10,20].

These sulfur compounds of unknown structure are highly instable and interconvert easily. This results in a lack of knowledge, to what extent these VSC-forming compounds are transferred when bottling the wine. The common addition of copper (II) salts to remove reductive (sulfidic) off-odors may even make the matter worse [17,21]. The influence of oxygen, whose ingress is highly dependent on the kind of closure, creates additional complications [22].

Appropriate strategies for managing wines with sulfurous off-aromas have been proposed [2]. The practicability of its application in "normal"-sized wine producing units has to be questioned.

This article is focused on those factors involved in the formation of polythionates and their reactions to elemental sulfur in different modifications, sulfur dioxide and H_2S. The effect of addition of copper (II) salts is also discussed. A later section will try to interpret contradicting and non-explainable findings under the assumption of polythionates being responsible for these effects.

Table 1. Presently-known sulfur rings and chains (n represents the chain length or ring size).

Type of Compound	Formula	(n *)	(n +)	Relevance in Wine
Sulfur homocycles	S_n	6–20	−80	Added as fungicide
(Poly-)sulfane	$H-S_n-H$	1–8	35	Component (H_2S_3) [23] Me_2S_3 [24–28]
Diorganopoly sulfane	$R-S_n-R$	1–7	35	Component disulfides ($n = 2$) [2]
Organopoly sulfides	$R-S_n-H$	1	2	Most likely i.e., persulfides ($n = 2$)
Polythionates	$^-O_3S-S_n-SO_3^-$	1–4	22	Most likely: $SO_2 + H_2S$ [29]
"Bunte salts"	$R-S_n-SO_3^-$	1	2–13	Most likely
Polythionates	$-S-S_n-SO_3^-$	0	1	Salts, i.e., sodium thiosulfate ($n = 0$)

n^*: in pure compounds, n^+: from spectroscopic or other evidence.

3. The Role of Elemental Sulfur

Although sulfur is one of the oldest known elements and in use since ancient times, the elucidation of its chemistry remains an ongoing task [30,31]. Among all elements, sulfur is the one with the most allotropes. The oxidation states of sulfur span eight numbers, from -2 in sulfides to $+6$ in sulfates. Elemental sulfur, often considered as chemically rather inert, reacts with many wine components. Sulfur is attacked by nucleophiles like thiols (thiolysis) or hydrogen sulfite (SO_2, sulfitolysis) due to the electrophilic character of -S-S-bonds [2]. It has been known for a long time that wine compounds with S-S groups could be formed from elemental sulfur residues used as pesticides [4–6]. The addition of sulfur prior to fermentation increases the formation of H_2S and even of the aroma thiol 3-mercaptohexanol (3-MH) [32]. Recent work has demonstrated that wines fermented in the presence of sulfur continue developing H_2S during storage. It was argued that these latent precursors could hardly arise from sulfur residues, due to its poor solubility in aqueous systems ($5 \ \mu g \ L^{-1}$ in water at 25 °C) and the fact that settling and racking can remove 95%.

Some speculations make the formation of polysulfanes responsible for the abovementioned effect. Indeed, homogenous elemental sulfur suspensions in aqueous organic mixtures on standing contain after some time polythionates, as outlined in the following section. Additionally, it should not be forgotten that elemental sulfur not only occurs in a water-insoluble form, but also as hydrophobic and hydrophilic nanoparticles [30,31,33]. The reactivity of these elemental sulfur nanoparticles among other forms of elemental sulfur is quite different from the crystalline modification. The variation of their surface areas, character and coatings reflect their analytical and physical-chemical properties and play an important role in geo- and bio-chemical processes involved in sulfur recycling.

Elemental sulfur is found as a component in must and wine, either as a residue from fungicidal treatment or produced by yeasts or bacteria ("biosulfur"). The latter has distinctly different properties from crystalline elemental sulfur. Its hydrophilic properties are the most striking of these differences, making it dispersible or even soluble in water [34], whereas crystalline inorganic sulfur is hydrophobic and will not be wetted by an aqueous solution. Nanoparticulate sulfur passes filters with mesh sizes of a few microns and less, usually used in wine processing.

Explanations for the hydrophilicity of extracellularly-stored sulfur globules, i.e., those produced by *Acidithiobacillus ferrooxidans*, can probably be given by the vesicle structure consisting mainly of polythionates (^-O_3S-S_n-SO_3^-). Due to the small particle size and hydrophilic surface, biologically-produced sulfur has advantages over sulfur flower in bioleaching and fertilizer applications. In a model proposed by Steudel et al. [35], the globules consist of a nucleus of sulfur rings (S_8) with water around it and long-chain sulfur compounds such as polysulfides or polythionates as amphiphilic compounds at the surface (Figure 1). Therefore, the following section deals with the chemistry of polythionates.

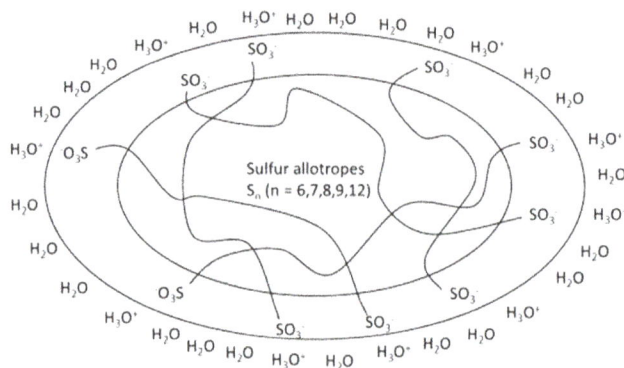

Figure 1. Schematic model for the composition of the particles in hydrophilic sulfur sols consisting of long-chain polythionates $S_n(SO_3^-)_2$ and sulfur homocycles S_n (according to [30,35]).

4. Polythionates and Polydithionates

The previous section about the role of elemental sulfur supports the theory that polythionates may play an important role in the chemistry of sulfur and its compounds in wine and must.

This class of sulfur chain-containing compounds may have been overlooked. They are closely related to polysulfanes and organopolysulfanes, which in some recent publications were made responsible for the reappearance of reductive off-odors [2,36,37]. In a broader context, polythionates may be considered as a subclass of polysulfanes due to their common structure with sulfur-sulfur bonds. Elemental sulfur, as well, should be treated in this connection, also because of its role in forming these compounds and their similar chemical reactivity due to the -S-S-bonds. In Table 1, the different types of compounds with sulfur rings and chains and their relation to wine chemistry are summarized.

5. History

The existence of polythionic acids (polythionates) dates back to the studies of John Dalton in 1808, devoted to the behavior of H_2S in aqueous solutions of SO_2. He found out that when introducing gaseous hydrogen sulfide into a freshly-prepared aqueous solution of SO_2, the smell of both gases vanishes and a milky liquid is formed [38]. Later on, this liquid was named "Wackenroder solution" after Heinrich Wilhelm Ferdinand Wackenroder, who conducted a systematic study (1846) [39]. Over the next 60–80 years, numerous studies showed the presence of ions, in particular tetrathionate and pentathionate anion ($S_4O_6{}^{2-}$ and $S_5O_6{}^{2-}$) respectively [40].

The mechanism of the formation of the "Wackenroder solution" still is not fully understood [41,42]. The reaction of hydrogen sulfide with sulfur dioxide in an aqueous solution yields as the primary product sulfoxylic acid ($H_2S_2O_2$), which reacts with further SO_2 to trithionate ($H_2S_3O_4$). By incorporation of sulfur, polythionates ($S_xO_6{}^{2-}$: $x = 4$, 5 and 6) and thiosulfate are formed [43]. Due to sulfitolysis, an equilibrium between these polythionates is formed. According to Schmidt [44], elemental sulfur reacts with SO_2 under sulfitolysis of the sulfur ring, which is finally degraded to thiosulfuric acid. Since free thiosulfuric acid is instable and decomposes to elemental sulfur, SO_2 and H_2S, the whole procedure will start de novo, but also, polythionates of a higher stability are formed.

5.1. Polythionates in Wine as Precursors

Recently, a polythionate (tetrathionate) was detected in wine [29]. Although this appearance is a first proof for speculations that organopolysulfanes and polysulfides could undergo two sulfitolysis steps to form polythionates, the reason why higher polythionates could not be detected remains unclear. To support this theory, tetrathionate could only be detected in sulfur-treated wine, but not in non-fermented model wine.

The majority of H_2S produced by yeast during wine fermentation is from the sulfate assimilation pathway (SAP), where sulfate is progressively reduced to sulfide, the precursor of the sulfur-containing amino acids cysteine and methionine, which are required for yeast growth, as outlined in Figure 2 [45].

Grape juice usually contains plenty of sulfate (~160–700 mg L^{-1}), but very low concentrations of cysteine and methionine (<20 mg L^{-1}), and therefore, the SAP is triggered during alcoholic fermentation to support yeast growth [5,46]. The mechanisms by which H_2S is released from the SAP are well studied and reviewed [5,47–49].

Handbooks for enology normally claim that excess H_2S is purged by the developing carbon dioxide (about 90%). Residues of high volatile sulfur-containing compounds (MeSH, EtSH) could also be removed by purging with inert gases like nitrogen or argon. However, what if these off-odor gases are trapped by SO_2 (employed for microbial must stabilization) or elemental sulfur as non-volatile, polar polythionates? These remain in wine and are transferred during bottling as latent sources for the reappearance of reductive off-odors.

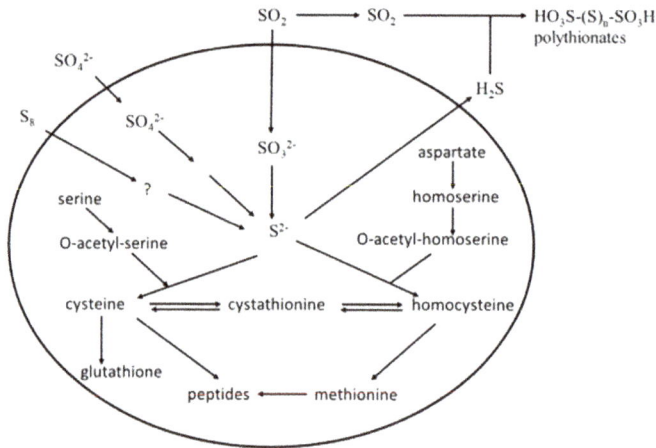

Figure 2. Sulfate assimilation pathway (SAP) in *Saccharomyces cerevisiae*. Top right: the formation of polythionates from the yeast metabolites H_2S and SO_2 ([20], modified).

Now, back to "real wine": As outlined above, elemental sulfur is used as a fungicide and may be transferred to the must. Must is stabilized by the addition of SO_2. Therefore, the formation of a "Wackenroder solution" (WS) is very likely. During fermentation, yeast produces H_2S in order to fulfil the need for the sulfur-containing amino acids methionine and cysteine via the sulfate assimilation pathway, simulating the original conditions for the formation of the "WS", containing polysulfanmono and disulfonic acids (polythionates). The experiments cited in [29] ran with 50 mg potassium bisulfite $K_2S_2O_5$ (which equals 33.3 mg SO_2) and 0–100 mg addition of elemental sulfur. In these experiments, up to 50 µg of total H_2S was formed (most in TECP = triscarboxyethyl phosphine releasable form), which is far over its odor threshold of 1 µg L^{-1}. This clearly demonstrates that if only 0.1% of the added sulfur reacts to polythionates, there is enough latent sources to form reductive off-odors.

Therefore, polythionates fulfil the criteria for the until now undefined precursors for reductive off-odors by Kreitman [2]: "They are metastabile during typical bottle storage conditions, i.e., the conversion of precursor to free sulfhydryl should occur on the order of several weeks to a couple of years at room temperature and their concentration should be large enough to generate concentrations of H_2S and MeSH (up to 1–2 $\mu M\ L^{-1}$)". In addition, they are polar compounds, highly soluble in aqueous solutions like wine, passing every filter system used in wine making.

5.2. The Role of Copper

A main topic in recent literature about the reappearance of reductive off-odors deals with the application of copper (II) salts [17,21,50–53]. The reason why copper fining removes or reduces sulfur-containing off-odor components is based on the very low solubility of copper sulfides and mercaptides in aqueous solutions, leading to their precipitation and the possibility of physical separation from the rest of the wine.

Copper plays an important role in winemaking practice, but its application is more and more questioned. Presently, trends try to minimize the amount of copper; even a total ban for applications of copper in vineyards is intended [54,55]. Many investigations in recent time showed that the common practice to add copper (II), either as sulfate or citrate, to remove sulfidic off-odors even may result in an elevated concentration of VSCs in the finished (bottled) wine.

Residual copper in white wine has been linked to oxidative and reductive spoilage processes, in haze formation (copper casse) and protein instability. More recent concerns include the coexistence of residual copper and hydrogen sulfide in wine stored under low oxygen conditions.

Copper-(II) salts (added in order to remove sulfidic off-odors) may react in a manifold way with all kinds of wine components, as outlined in Figure 3.

The chemistry of copper in white and red wine does not differ considerably. There are some overlapping issues, especially considering sulfidic off-odors. However, due to the quite different phenolic status of red wine, this review deals with the state of knowledge of copper in white wine.

Explanations of the reaction scheme (Figure 3) [56] are given below:

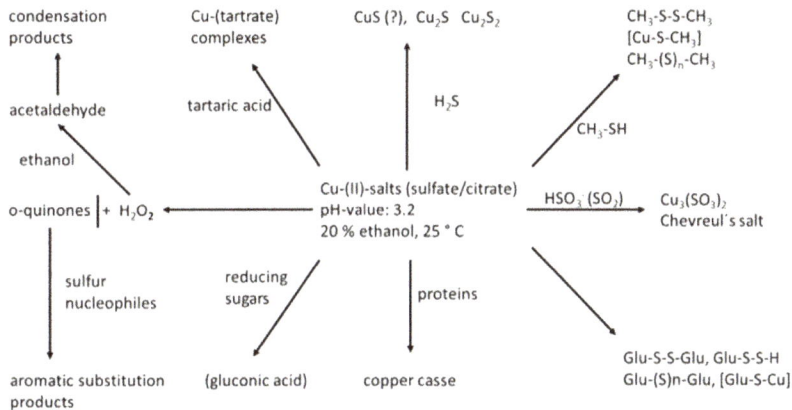

Figure 3. Reactions of wine components with copper ([56], modified).

(1) On reaction with H_2S, no simple precipitation of insoluble CuS takes place; the "copper sulfide" consists of a mixture of Cu_2S and CuS_2, better described as $[Cu_3S_3]$. In addition, di- and poly-sulfanes are formed [29,53].

(2) Reaction with "off-odor" VSCs MeSH and EtSH leads to the formation of disulfides, mixed disulfides [51], including those with H2S, which leads to hydropersulfides (R-S-S-H), the formation of dialkylpolysulfanes and alkylpolysulfanes, as well as mixed forms with organyl residues (R_1-(S) n-R_2); the formation of copper-(I)-mercaptides and soluble complexed forms of these [53].

(3) If wines are treated with copper (II) salts alone, almost the total disappearance of thiols like 3-MH and the concomitant formation of disulfides were observed [51].

(4) Reaction with sulfur dioxide (rather, bisulfite): There are ideal reaction conditions to form "Chevreul's salt" $[Cu_3(SO_3)_2]$, a highly stable mixed valence sulfite compound [57,58]. Because of its high insolubility, the precipitation of Chevreul's salt is used in hydrometallurgy.

(5) Glutathione behaves like a normal thiol in forming the di- and polysulfanes and the corresponding mercaptide. This has to be considered in the light of using glutathione in enology (OIV) [59,60].

(6) Glucose (reducing aldoses): Copper (II) sulfate is used for the quantitative determination of reducible sugars in wine analytics according to Rebelein. Although this method works in strong alkaline solution (the aldehyde group is oxidized to a carboxyl group; glucose is converted to gluconic acid), there still might be some evidence that copper (II) is converted to Cu (I) at normal wine pH of around 3.0–3.5.

(7) Copper addition also induces the formation of hazes in wine by reaction with proteins.

(8) Copper, especially in combination with iron ions, catalyzes the oxidation of wine o-diphenols and other phenolics like caftaric acid, cyanidin, catechin, epicatechin, gallic acid and its derivatives, to form highly reactive o-quinones. The impact on sulfur-containing aroma active compounds lies in their high nucleophilicity, giving rise to irreversible 1,4-Michael additions [61].

(9) o-Quinones are also made responsible for the development of volatile sulfur compounds by the Strecker reaction with cysteine and methionine [9].

(10) Copper forms stable complexes with tartaric acid [17,21,62].

Copper may be present in grape juice due to its application as a fungicide in the vineyard. Copper concentrations in must and wine range from 0.2 mg/L–7.3 mg/L. In a recent survey of 100 Chardonnay juices during the 2009 Australian vintage, the median Cu concentration was around 1 mg/L with the lowest at 0.2 and the highest at 7.0 mg/L. In a recent publication, different forms of metals in wine were claimed. Thus, a fractionation of Cu and Fe compounds into hydrophobic, cationic and residual forms was described [61]. Considering wine components, which react with copper ions, forming insoluble compounds like sulfidic species or the mixed valence "Chevreul's salt" with sulfur dioxide, the question may be raised about how these species will contribute to the different fractions.

5.3. Some Remarks on the Oxidation States of Copper/Sulfur Compounds

The bonding in copper sulfides cannot be correctly described in terms of a simple oxidation state formalism because the Cu-S bonds are somewhat covalent rather than ionic in character and have a high degree of delocalization, resulting in complicated electronic band structures. Many textbooks (e.g., [63]) give preference to the mixed valence formula $(Cu^+)_2(Cu^{2+})(S^{2-})(S_2)^{2-}$ For CuS, X-ray photoelectron spectroscopic data give strong evidence that, in terms of the simple oxidation state formalism, all the known copper sulfides should be considered as purely monovalent copper compounds, and a more appropriate formulae would be $(Cu^+)_3(S^{2-})(S_2)^-$ for CuS and $(Cu^+)(S_2)^-$ for CuS$_2$, respectively. An explanation therefore is given by the fact that the highly polarizing Cu^{2+}-ion removes an electron from the highly polarizable S^{2-}-anion:

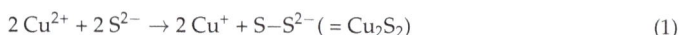

$$2\,Cu^{2+} + 2\,S^{2-} \rightarrow 2\,Cu^+ + S{-}S^{2-}\,(= Cu_2S_2) \tag{1}$$

Copper (II) reacts with thiols by reduction to copper (I) and the formation of disulfides and copper (I)-mercaptides or related complexes (Equation 2). Both are suspected to be sources for the reappearance of reductive off-odors during bottle-aging [17,21,53].

$$4\,Cu^{2+} + 4\,R\text{-}S\text{-}H \rightarrow 2\,Cu^+ + R\text{-}S\text{-}S\text{-}R\ 2\,[R\text{-}S\text{-}Cu] \tag{2}$$

Compounds with the co-existence of divalent Cu^{2+}-ions in sulfidic compounds with oxidation state −2 simply do not exist.

The same conclusions may be drawn for the co-existence of copper (II) and SO$_2$ in wine and must. Hydrogen sulfite ions, the most probable species at wine pH, react with copper (II) salts to form Chevreul's salt, a mixed valence sulfite of copper Cu$_3$(SO$_3$)$_2$ better represented by the formula [64]:

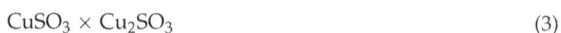

$$CuSO_3 \times Cu_2SO_3 \tag{3}$$

In Chevreul's salt crystals, there are two environments for copper. The +1 oxidation state copper is in a distorted tetrahedral space surrounded by three oxygens and a sulfur atom. The +2 oxidation state copper is in a distorted octahedral coordination surrounded by four oxygen atoms and two water molecules, which stresses the above statement [65]. Supposing must or wine contain average concentrations of 32 mg (0.5 mMol) free SO$_2$ and 5 mg (about 0.1 mMol) copper (II) ions, the formation of Chevreul's salt is very likely.

5.4. The Role of Sulfur Dioxide

Sulfur dioxide SO$_2$ is the most widely-used additive in winemaking. Its antimicrobial effect, antioxidant properties and its capability to protect wine from various unwanted reactions makes it an unavoidable additive in winemaking [59]. The legal limit for total SO_2 concentration according to the OIV is 150–200 mg L^{-1} for dry wines, while in exceptional cases, it can reach up to 400 mg L^{-1}.

Sulfur dioxide participates in the formation of reductive sulfurous off-odors (Figure 4).

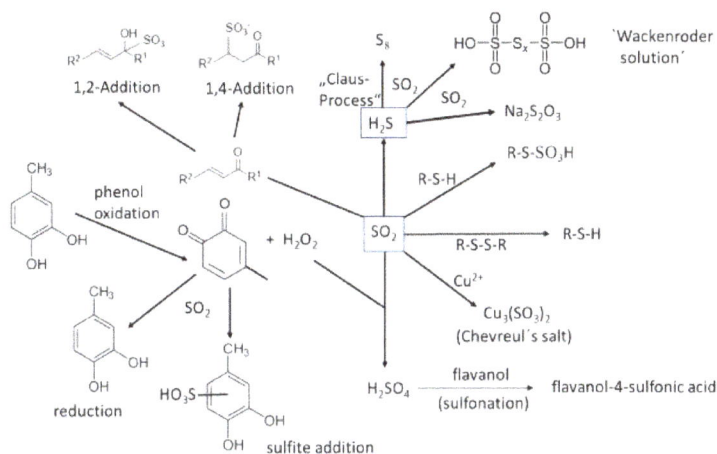

Figure 4. Reactions of SO_2 with wine components (top right: the formation of polythionates from the yeast metabolites H_2S and SO_2).

Besides the direct enzymatic reduction to hydrogen sulfide in the sulfate assimilation pathway (Figure 2), numerous different chemical pathways exist in a very complex system that makes it almost impossible to follow reductive odor compounds analytically.

In recent studies, SO_2 in combination with copper (II) salts was cited as being responsible for the development of reductive off-odors [65]. Surprisingly, this effect was only observed in "real" wines, not in so-called model wines, usually employed in these trials. An explanation for this effect may easily be found in the circumstance that model wine has not passed a fermentation step. Therefore, all yeast-born "sulfurous compounds" (including polythionates) are simply not present and thus cannot result various VSCs by treatment with copper.

The following list refers to Figure 4, explaining some of the represented reactions.

(1) The oxidation of wine o-diphenols leads to o-quinones and hydrogen peroxide, which oxidize sulfite to sulfuric acid. Highly reactive aromatic compounds like phenols, indoles, etc., may undergo electrophilic aromatic substitution (sulfonation) [66].

(2) SO_2 (as HSO_3^-) itself is a strong nucleophile, which reacts with o-quinones in a Michael addition-type reaction under the irreversible formation of aromatic sulfonic acids [61].

(3) SO_2 reacts with various carbonyl compounds forming bisulfite-addition products ("bound SO_2"). α,β-unsaturated ketones (i.e., the wine-relevant aroma compound β-damascenone) form 1,4-addition products [67].

(4) At elevated temperatures, H_2S and SO_2 comproportionate to elemental sulfur. This reaction is used in the technical "Claus-process" for desulfuration of crude gas.

(5) At ambient temperature, H_2S and SO_2 react to the so-called "Wackenroder solution", a mixture of polythionates, H_2S, SO_2 and elemental sulfur (see Section 5).

(6) The formation of (at wine pH) instable thiosulfuric acid takes also part in the formation of "Wackenroder solution".

(7) Thiols are powerful reducing agents that may be oxidized by sulfur (IV) compounds to mixtures of polysulfanes. The first step of this reaction is the formation of S-sulfonated thiols.

(8) Recently, various sulfonated compounds were observed in wine for the first time (e.g., S-sulfonated cysteine and glutathione) [66].

(9) The addition of SO_2 to anthocyanins results in colorless 4-sulfonic acids.

(10) The addition of copper (II) to sulfite-containing solutions leads to the formation of a mixed valence copper sulfite ("Chevreul's salt") [68].

(11) SO_2 may cleave disulfide bonds under the formation of thiols ("sulfitolysis") [69].

(12) The so-called "sulfite degradation" of sulfur chains in elemental sulfur, polysulfanes and polythionates is part of the very complicated sulfur chemistry in aqueous systems (including wine).

6. Discussion of Polydithionates: The Answer to Some Unsolved Questions?

This section starts with the assumption that polythionates are part of the non-volatile and odorless forms in wine, acting as precursors for the redevelopment of reductive sulfurous off-odors. The following findings and statements mainly made by Ferreira and Franco-Luesma in the literature [70] may then be seen in a clearer light.

(1) Wine contains H_2S and MeSH in non-volatile and odorless forms. An analytical method has been developed to detect H_2S and MeSH present in bonded forms [71]. We claim polythionates to be part of the non-volatile and odorless forms.

(2) Ferreira and Franco-Luesma [70] pointed out that each wine has a specific ability to bind reversibly H_2S and mercaptans. "If free H_2S and MeSH are added to different wines, some of these molecules remain volatile, and hence can be easily smelled, while there are some others in which the same molecules are so strongly bonded that they cannot be measured (and hence smelled) as free forms" [70] (bonded forms include polythionates; → see Daltons' experiment [38]). The reversible binding of H_2S in polythionates lies in the labile character of these compounds. They easily decay into H_2S, SO_2 and elemental sulfur.

(3) Oxygen only reacts in converting H_2S and mercaptans into their free forms. Bound or complexed forms (including polythionates) require longer periods of treatment to release the free forms [37].

(4) Even under strict oxygen-free conditions, H_2S is able to slowly, but irreversibly (?) react to something (we claim polythionates) in some wine.

(5) The increments of free H_2S observed when wine is stored in anoxia are mostly the release of complexed forms: This is valid for polythionates and other components of the "Wackenroder solution", if considered as part of the "complexed forms" [29].

(6) Analytical methods to detect H_2S, MeSH and other VSCs in the headspace do not work, if they suffer interactions in wines occasionally so strong that in some wines, they are not released to the wine headspace. In contrast, a previously-developed method in which the wine is strongly diluted with brine revealed that such interactions are reversible, demonstrating that wines contain a large pool of reversible complexed (as polythionates) non-volatile forms of H_2S and mercaptans [72].

(7) TCEP (Tris (2-carboxyethyl) phosphine) was evaluated for its ability to release complexed H_2S from wine. Surprisingly, even though TCEP was much less effective than brine at releasing H_2S from copper sulfide complexes in the recovery experiments, higher concentrations of H_2S released by TCEP were observed in six of the seven wines (up to 25.3 µg/L). These results suggest the presence of additional TCEP-releasable sources (polythionates) of H_2S (or thiol interferences) in wine [48]. Since TCEP acts specifically to break -S-S-bonds, polythionates will also be attacked with the release of H_2S.

(8) Elemental sulfur pesticide residues on grapes can not only produce H_2S during fermentation, but are also able to form precursors capable of generating additional H_2S after bottle storage. Among these precursors, glutathione tri- and polysulfanes ($Glu\text{-}S\text{-}S_n\text{-}S\text{-}Glu$) have been detected, but also for the first time, tetrathionate ($S_4O_6^{2-}$). These precursors are polar, wine-soluble intermediates that can be converted into H_2S following TCEP (triscarboxyethyl phosphine)

addition [29]. The identity of these intermediates was unclear, but could potentially involve polysulfide (polythionates) adducts. The authors claim sulfitolysis of polysulfanes as the reason; the classical formation by the "Wackenroder reaction" from the wine components H_2S and SO_2 is not mentioned. These findings come closest to our "polythionate theory" made in (6).

(9)   Wine producers know that if a wine has a tendency to develop or had reductive off-odors, it is possible that the off-flavor is reoccurring during the storage of the bottle, in particular if closures with low permeability to oxygen are used. This indicates that the oenological treatments affect the odor-active compounds, but not their diverse precursors [56,70]. Polythionates might belong to this type of precursor.

(10)   Up to this date, these precursors have not been identified. One reason why the progress in research has been so difficult lies in the complicated chemistry of sulfur. This element can be present in different redox states, has the ability to concatenate (polysulfanes and polythionic acids), making bonds with itself, and can form a large number of combinations [16]. These statements support our polythionates theory.

(11)   Copper, acting on unknown precursors, was associated with large increases of H_2S in Shiraz wines [73]. The unknown precursors may be polythionates born in the fermentation step. There is no remarkable increase in model wine when treated with copper, due to the lack of these precursors.

(12)   This suggests that a pool of yet to be identified precursor compounds (polythionates) is also involved in modulating final concentrations in wine through copper-catalyzed reactions [74] (since copper salts are active in cleaving -S-S-bonds, these bonds will be cleaved in polythionates, as well).

The comparison of experimental results obtained in real wine with those obtained in model wine should be taken with care. Since model wine has not passed a fermentation step, it lacks all the yeast-born precursors (i.e., polythionates).

7. Conclusions

Still other latent sources of H_2S and MeSH may exist, like the potential contribution of non-volatile asymmetric disulfides, trisulfides or even polysulfanes (e.g., adducts with other thiol compounds), which have not been well explored. Since the natural tripeptide glutathione [75,76] represents the thiol with the highest concentration in must and wine (up to 120 mg), this sulfanyl compound is strongly involved in these processes of reductive off-odor formation [16,29]. This topic would need its own section.

Among these compounds with molecules containing sulfur chains, polythionates have not been taken into consideration, although their existence has been known since the early development of modern chemistry. The combination of added or biologically-produced SO_2 and H_2S in must and wine form an ideal basis for their formation. Although it seems that the existence of these species can give more explanations for some strange findings in the matter of the reappearance of reductive odors in wine, these elucidations neither facilitate the whole complex, nor give a clear mode of action for how to quantify and remove these off-odors. Polythionates are part of the mystery that still exists about sulfur chemistry. Their instability, decay and de novo formation, as well as their interconvertibility make it almost impossible to establish a reliable analytical method.

The addition of copper to remove reductive off-odors even worsens the situation, since little is known about the interactions of copper in both oxidation states with these species. The presence of copper may lead to an immediate decrease of smelling free forms of H_2S and MeSH, but the danger from a reservoir of unknown precursors for reductive off-odors is transferred into the bottles, with a high likelihood of reappearance during storing.

Numerous different chemical pathways exist in a very complex system that would make it almost impossible to follow reductive odor compounds analytically. The suspicious mixture of unknown instable sulfur intermediates may now give an explanation for the reappearance of sulfur-caused reductive off-odors, but hardly can recommend a mode of action. Furthermore, it is clearly demonstrated that the addition of copper (II) rather creates more confusion than helping to solve the problem.

We hope to have given part of an answer to the question and the statement made by Waterhouse et al. in his recently published textbook on wine chemistry [16]: "Finally, other potent (and common) contributors to reduced aromas may have been overlooked. If the standing hypotheses are not valid, then these questions are founded on unknown chemical processes; that is, what "reducing" agents or reactions are participating or occurring? Wine storage is a rare situation of very long lifetimes of reactive solutions where kinetically slow processes, perhaps those not commonly observed otherwise, have time to occur.

Determining the likely latent sources of H_2S, CH_3SH, or other causes of reduced aromas is more than interesting chemistry—it is also a crucial step in eliminating an increasingly vexing problem. Determining their identity should lead to better prevention, remediation, and detection strategies for reduced aroma formation in bottle. Currently, some winemakers will add ascorbic acid to predict the appearance of reduced aromas, but the rationale for this test is based on disulfides as latent precursors, and these tests have not been validated. In recent years, winemakers have benefited from accelerated aging tests, for example, contact tests to predict the likelihood of potassium bitartrate instability and it is expected that a validated accelerated reduction assay would be equally beneficial" [16].

Author Contributions: N.M. outlined the review and wrote the biggest part of the review. He realized by intensive literature study, that the formation of polythionates, hitherto not taken into consideration, could be the reason for the reappearance of sulfurous off-odours. D.R. gave intensive support in designing the review and by numerous technical discussions. Both authors contributed equally in revision and final version of the manuscript.

funding: This research received no external funding.

Conflicts of Interest: The authors declare no conflict of interest.

References

1. Goode, J.; Harrop, S. Wine faults and their prevalence: Data from the world's largest blind tasting. In Proceedings of the Les XXes Entretiens Scientifiques Lallemand, Horsens, Denmark, 15 May 2008; pp. 7–9.
2. Kreitman, G.Y.; Elias, R.J.; Jeffery, D.W.; Sacks, G.L. Loss and formation of malodourous volatile sulfhydryl compounds during wine storage. *Crit. Rev. Food Sci Nutr.* **2018**, *5*, 1–24. [CrossRef] [PubMed]
3. Jiranek, V.; Langridge, P.; Henschke, P.A. Regulation of hydrogen sulfide liberation in wine producing. Saccharomyces cerevisiae by assimilable nitrogen. *Appl. Environ. Microbiol.* **1995**, *61*, 461–467. [PubMed]
4. Rauhut, D.; Kürbel, H. The production from H_2S from elemental sulfur residues during fermentation and its influence on the formation of sulfur metabolites causing off-flavours in wine. *Vitic. Enol. Sci.* **1994**, *49*, 27–36.
5. Rauhut, D. Usage and formation of sulfur compounds. In *Biology of Microorganisms on Grapes, in Must and in Wine*; König, H., Unden, G., Fröhlich, J., Eds.; Springer International Publishing AG: Cham, Switzerland, 2017; pp. 255–291.
6. Rauhut, D. Yeast-Production of sulfur compounds. In *Wine Microbiology and Biotechnology*; Fleet, G.H., Ed.; Harwood Academic Publishers: Chur, Switzerland, 1993; pp. 183–223.
7. Huang, C.-W.; Walker, M.E.; Fredrizzi, B.; Gardner, J.V. Hydrogen sulfide and its roles in Saccharomyces cerevisiae in wine making context. *FEMS Yeast Res.* **2017**, *17*, fox058. [CrossRef] [PubMed]
8. Rauhut, D.; Kürbel, H.; Dittrich, H.H.; Grossmann, M. Properties and differences of commercial yeast strains with respect to their formation of sulfur compounds. *Vitic. Enol. Sci.* **1996**, *51*, 187–192.
9. Pripis-Nicolau, L.; de Revel, G.; Bertrand, A.; Maujean, A. Formation of Flavor Components by the Reaction of Amino Acid and Carbonyl Compounds under Mild Conditions. *J. Agric. Food Chem.* **2000**, *48*, 3761–3766. [CrossRef] [PubMed]
10. Müller, N. Thiaminpyrophosphat-ein natürlich vorkommendes Iminiumsalz. *Z. Naturforsch. B* **2014**, *69*, 489–500. [CrossRef]
11. Lloyd, D. Hydrogen sulfide: Clandestine microbial messenger? *TRENDS Microbiol.* **2006**, *14*, 457–462. [CrossRef] [PubMed]

12. Lloyd, D.; Murray, D.B. The temporal architecture of eukaryotic growth. *FEBS Lett.* **2006**, *580*, 2830–2835. [CrossRef] [PubMed]

13. Reschke, S.; Tran, T.; Bekker, M.; Wilkes, E.; Johnson, D. Using copper more effectively in winemaking. *Wine Vitic. J.* **2015**, *9*, 35–39.

14. Görtges, S. Böckserbeseitigung mit Kupfercitrat. *Der Deutsche Weinba* **2009**, *5*, 24–25.

15. Steidl, R. Der Einsatz von Silberchlorid zur Böckserbekämpfung. *Mitt. Klosterneubg.* **2010**, *3*, 92–95.

16. Waterhouse, A.L.; Sacks, G.L.; Jeffery, D.W. (Eds.) Appearance of Reduced Aromas during Bottle Storage. In *Understanding Wine Chemistry*; John Wiley & Sons: Chichester, UK, 2016; pp. 397–399.

17. Clark, A.C.; Wilkes, E.N.; Scollary, G.R. Chemistry of Copper in white wine: A. review. *Aust. J. Grape Wine Res.* **2015**, *21*, 339–350. [CrossRef]

18. Kinzurik, M.I.; Herbst-Johnstone, M.; Gardner, R.C.; Fedrizzi, B. Evolution of Volatile Sulfur Compounds during Wine Fermentation. *J. Agric. Food Chem.* **2015**, *63*, 8017–8024. [CrossRef] [PubMed]

19. Kinzurik, M.I.; Herbst-Johnstone, M.; Gardner, R.C.; Fedrizzi, B. Hydrogen sulfide production during yeast fermentation causes the accumulation of ethanethiol, S-ethyl thioacetate and diethyl disulfide. *Food Chem.* **2016**, *209*, 341–347. [CrossRef] [PubMed]

20. Müller, N. Iminiumsalz-Strukturen bei der durch Pyridoxalphosphat (Vitamin B6) katalysierten Bildung von Aromastoffen und Fehlaromen im Wein. *Z. Naturforsch. B* **2018**, *73*, 521–533. [CrossRef]

21. Clark, A.C.; Grant-Peerce, P.; Cleghorn, N.; Scollary, G.R. Copper (II) addition to white wines containing hydrogen sulfide: Residual copper concentration and activity. *Aust. J. Grape Wine Res.* **2015**, *21*, 30–39. [CrossRef]

22. Bekker, M.Z.; Day, M.P.; Holt, H.; Wilkes, E.; Smith, P.A. Effect of oxygen exposure during fermentation on volatile sulfur compounds in Shiraz wine and a comparison of strategies for remediation of reductive character. *Aust. J. Grape Wine Res.* **2016**, *22*, 24–35. [CrossRef]

23. Starkenmann, C.; Chappuis, C.J.-F.; Niclass, Y.; Deneulin, P. Identification of Hydrogen Disulfanes and Hydrogen Trisulfanes in H2S Bottle, in Flint, and in Dry Mineral White Wine. *J. Agric. Food Chem.* **2016**, *64*, 9033–9044. [CrossRef] [PubMed]

24. Nishibori, N.; Kuroda, H.; Yamada, O.; Goto-Yamamoto, N. Factors Affecting Dimethyl Trisulfide Formation in Wine. *Food Sci. Technol. Res.* **2017**, *23*, 241–248. [CrossRef]

25. Gijs, L.; Perpète, P.; Timmermans, A.; Collin, S. 3-Methylthiopropionaldehyde as Precursor of Dimethyl Trisulfide in Aged Beers. *J. Agric. Food Chem.* **2000**, *48*, 6196–6199. [CrossRef] [PubMed]

26. Fan, W.; Xu, Y. Characteristic Aroma Compounds of Chinese Dry Rice Wine by Gas-Chromatography-Olfactometry and Gas-Chromatography-Mass Spectroscopy. In *Flavor Chemistry of Wine and Other Alcoholic Beverages*; ACS Symposium Series 2012; American Chemical Society: Washington, DC, USA, 2012; Chapter 16; pp. 277–301.

27. Isogai, I.; Utsonomiya, H.; Kanda, R.; Iwata, H. Changes in Aroma Compounds of Sake during Aging. *J. Agric. Food Chem.* **2005**, *53*, 4118–4123. [CrossRef] [PubMed]

28. Landaud, S.; Helinck, S.; Bonnarme, P. Formation of volatile sulphur compounds and metabolism of methionine and other sulfur compounds in fermented food. *Appl. Microbiol. Biotechnol.* **2008**, *77*, 1191–1205. [CrossRef] [PubMed]

29. Jastrembski, J.A.; Allison, R.B.; Friedberg, E.; Sacks, G.L. Role of Elemental Sulfur in Forming Latent Precursors of H2S in Wine. *J. Agric. Food Chem.* **2017**, *65*, 10542–10549. [CrossRef] [PubMed]

30. Kafantaris, F.C. On the Reactivity of Nanoparticulate Elemental Sulfur: Experimentation and Field Observations. Ph.D. Thesis, Indiana University, Bloomington, IN, USA, December 2017.

31. Steudel, R.; Eckert, B. Solid sulfur allotropes. In *Elemental Sulfur and Sulfur-Rich Compounds*; Topics in Current Chemistry Book Series; Springer: Berlin/Heidelberg, Germany, 2003; Volume 230, 79p.

32. Araujo, L.D.; Vannevel, S.; Buica, A.; Callerot, S.; Fredrizzi, B.; Kilmartin, P.A.; Du Toit, W.J. Indications of the prominent role of elemental sulfur in the formation of the varietal thiol 3-mercapto- hexanol in Sauvignon blanc. *Food Res. Int.* **2017**, *98*, 79–86. [CrossRef] [PubMed]

33. Krahn, R. *Mechanismen der Schwefelwasserstofferzeugung durch Saccharomyces Cerevisiae für die Biotechnologische Immobilisierung von Schwermetallen*; Inaugural-Dissertation: Düsseldorf, Germany, 2000.

34. Kleinjan, W.E.; Keizer, A.; Janssen, A.J.H. Biologically Produced Sulfur. *Top. Curr. Chem.* **2003**, *230*, 167–188.

35. Steudel, R. Aqueous Sulfur Sols. *Top. Curr. Chem.* **2003**, *230*, 153–166.

36. Nedjma, M.; Hoffmann, N. Hydrogen sulfide reactivity with thiols in the presence of copper (II) in hydroalcoholic solutions. *J. Agric. Food Chem.* **1996**, *44*, 3935–3938. [CrossRef]
37. Vela, E.; Purificíon, H.-O.; Franco-Luesma, E.; Ferreira, V. Micro-oxygenation does not eliminate hydrogen sulfide and mercaptans from wine; it simply shifts redox and complex-related equilibria to reversible oxidized species and complexed forms. *Food Chem.* **2018**, *243*, 222–230. [CrossRef] [PubMed]
38. Dalton, J.; Wolff, F. *Ein neues System des Chemischen Theils der Naturwissenschaft*; Verlag Eduard Hitzig: Berlin, Germany, 1813; Volume 2, pp. 190–196.
39. Wackenroder, H. Über eine neue Säure des Schwefels. *Arch. Pharm.* **1846**, *97*, 272–288. [CrossRef]
40. Tomozo, K. Analytical Chemistry of Polythionates and Thiosulfate. A. Review. *Anal. Sci.* **1990**, *6*, 3–14. [CrossRef]
41. Blasius, E.; Burmeister, W. Untersuchung der Wackenroderschen Reaktion mit Hilfe der radiopapierchromatographischen Methode. *Fresenius' Z. Anal. Chem.* **1959**, *168*, 1–15. [CrossRef]
42. Stamm, H.; Seipold, O.; Goehring, M. Zur Kenntnis der Polythionsäuren und ihrer Bildung. 4. Mitteilung. Die Reaktionen zwischen Polythionsäuren und schwefliger Säure bzw. Thioschwefelsäure. *Z. Allg. Anorg. Chem.* **1941**, *247*, 93–98. [CrossRef]
43. Drozdova, Y.; Steudel, R. The Reaction of H_2S with SO_2: Molecular Structures, Energies, and Vibrational Data of Seven Isomeric Forms of H_2SO_3. *Chem. A Eur. J.* **1995**, *1*, 193–198. [CrossRef]
44. Schmidt, M.; Heinrich, H. Beitrag zur Lösung des Problems der Wackenroderschen Flüssigkeit. Über Säuren des Schwefels, XII. *Angew. Chem.* **1958**, *70*, 572–573.
45. Thomas, D.; Surdin-Kerjan, Y. Metabolism of Sulfur Amino Acids in *Saccharomyces cerevisiae*. *Mol. Biol. Rev.* **1997**, *61*, 503–532.
46. Ugliano, M.; Henschke, P.A. Yeasts and wine flavour. In *Wine Chemistry and Biochemistry*; Moreno-Arribas, M.V., Polo, M.C., Eds.; Springer: New York, NY, USA, 2009; pp. 313–392.
47. Swiegers, J.H.; Pretorius, I.S. Modulation of volatile sulfur compounds by wine yeast. *Appl. Microbiol. Biol.* **2007**, *74*, 954–960. [CrossRef] [PubMed]
48. Chen, Y.; Jastrzembski, J.A.; Sacks, G.L. Copper-Complexed Hydrogen Sulfide in Wine. Measurement by Gas Detection Tubes and Comparison of Release Approaches. *Am. J. Enol. Vitic.* **2017**, *68*, 191–199. [CrossRef]
49. Ugliano, M.; Kwiatkowski, M.; Vidal, S.; Capone, D.; Siebert, T.; Dieval, J.-B.; Aagard, O.; Waters, E.J. Evolution of 3-mercaptohexanol, hydrogen sulfide, and methyl mercaptan during bottle storage of Sauvignon blanc wines. Effect of glutathione, copper, oxygen exposure, and closure-derived oxygen. *J. Agric. Food Chem.* **2011**, *59*, 2564–2572. [CrossRef] [PubMed]
50. Vela, E.; Hernández-Orte, P.; Franco-Luesma, E.; Ferreira, V. The effects of copper fining on wine content in sulfur off-odors and on their evolution during accelerated anoxic storage. *Food Chem.* **2017**, *231*, 212–221. [CrossRef] [PubMed]
51. Roland, A.; Delpech, S.; Dagan, L.; Ducasse, M.-A.; Cavelier, F.; Schneider, R. Innovative Analysis of 3-mercaptohexan-1-ol, 3-mercaptohexylacetate and their corresponding disulfides in wine by stable isotope dilution assay and nano-liquid chromatography tandem mass spectrometry. *J. Chromatogr. A* **2016**, *1468*, 154–163. [CrossRef] [PubMed]
52. Kreitman, G.Y.; Danilewicz, J.C.; Jeffery, D.W.; Elias, R.J. Reaction Mechanisms of Metals with Hydrogen Sulfide and Thiols in Model Wine. Part 1: Copper-Catalyzed Oxidation. *J. Agric. Food Chem.* **2016**, *64*, 4095–4104. [CrossRef] [PubMed]
53. Kreitman, G.Y.; Danilewicz, J.C.; Jeffery, D.W.; Elias, R.J. Copper (II)-Mediated Hydrogen Sulfide and Thiol Oxidation to Disulfides and Organic Polysulfanes and Their Reductive Cleavage in in Wine. Mechanistic Elucidation and Potential Applications. *J. Agric. Food Chem.* **2017**, *65*, 2564–2571. [CrossRef] [PubMed]
54. Fuchs, R. *Künftig ohne Kupfer*; Heft 11/18; Der Deutsche Weinbau: Neustadt an der Weinstraße, Germany, 2018; pp. 16–18.
55. Altmayer, B. *Genehmigung auf der Kippe*; Heft 5/18; Das Deutsche Weinmagazin: Neustadt an der Weinstraße, Germany, 2018; pp. 32–35.
56. Müller, N.; Rauhut, D. Neuere Erkenntnisse zur Behandlung von reduktiven Noten und Böcksern im Wein. In *Deutsches Weinbau Jahrbuch*; Eugen Ulmer: Stuttgart, Germany, 2018; pp. 83–94.
57. Calban, T.; Sevim, F.; Lacin, O. Investigation of Precipitation Conditions of Chevreul's Salt. *Int. J. Chem. Mol. Eng.* **2016**, *10*, 1021–1024.
58. Pietsch, E.H.E. *Gmelin's Handbook*; Verlag Chemie: Weinheim, Germany, 1958; Volume 8, p. 484.

59. OIV. *International Code of Oenological Practices*; International Organisation of Vine and Wine: Pairs, France, 2015.

60. Ribéreau-Gayon, G.Y.; Maujean, A.; Dubourdieu, D. *Handbook of Enology: The Chemistry of Wine Stabization and Treatment*; John Wiley & Sons: Chichester, UK, 2006; Volume 2.

61. Nikolantonaki, M.; Chichuc, J.; Teissedre, P.-L.; Darriet, P. Reactivity of volatile thiols with polyphenols in a wine-model medium: Impact of oxygen, iron, and sulfur dioxide. *Anal. Chim. Acta* **2010**, *660*, 102–109. [CrossRef] [PubMed]

62. Rousseva, M.; Kontoudakis, N.; Schmidke, L.M.; Scollary, G.R.; Clark, A.C. Impact of wine production on the fractionation of copper and iron in Chardonnay wine: Implications for oxygen consumption. *Food Chem.* **2016**, *203*, 440–447. [CrossRef] [PubMed]

63. Greenwood, N.N.; Earnshaw, A. *Chemistry of the Elements*, 2nd ed.; Butterworth-Heinemann: Oxford, UK; Waltham, MA, USA, 1997; ISBN 0-08-037941-9.

64. Kierkegaard, P.; Nyberg, B. The crystal structure of $Cu_2SO_3 \cdot CuSO_3 \cdot 2H_2O$. *Acta Chem. Scand.* **1965**, *19*, 2189–2199. [CrossRef]

65. Bekker, M.Z.; Smith, M.E.; Smith, P.A.; Wilkes, E.N. Formation of Hydrogen Sulfide in Wine: Interactions between Copper and Sulfur Dioxide. *Molecules* **2016**, *21*, 1214. [CrossRef] [PubMed]

66. Arapitsas, P.; Guella, G.; Mattivi, F. The impact of SO_2 on wine flavanols and indoles in relation to wine style and age. *Sci. Rep.* **2018**, *25*, 858. [CrossRef] [PubMed]

67. Duhamel, N.; Piano, F.; Davidson, S.J.; Larcher, R.; Fredrizzi, B.; Barker, D. Synthesis of alkyl sulfonic acid aldehydes and alcohols, putative precursors to important wine aroma thiols. *Tetrahedron Lett.* **2015**, *56*, 1728–1731. [CrossRef]

68. Chevreul, M.E. Propriétés du sulfite de cuivre. *Anal. Chim.* **1812**, *83*, 187–190.

69. Bobet, R.A.; Noble, A.C.; Boulton, R.B. Kinetics of the ethanethiol and diethyl disulfide interconversion in wine like solutions. *J. Agric. Food Chem.* **1990**, *38*, 449–452. [CrossRef]

70. Ferreira, V.; Franco-Luesma, E. Understanding and Managing Reduction Problems. *Int. J. Enol. Vitic.* **2016**, *2*, 1–13.

71. Franco-Luesma, E.; Ferreira, V. Quantitative analysis of free and bonded forms of volatile sulfur compounds in wine. Basic methodologies and evidences showing the existence of reversible cation-complexed forms. *J. Chromatogr. A* **2014**, *1359*, 8–15. [CrossRef] [PubMed]

72. Franco-Luesma, E.; Ferreira, V. Reductive off-odours in wine: Formation and release of H_2S and methanethiol during the accelerated anoxic storage of wines. *Food Chem.* **2016**, *199*, 42–50. [CrossRef] [PubMed]

73. Bekker, M.Z.; Mierczynska-Vasilev, A.; Smith, P.; Wilkes, E.N. The effects of pH and copper on the formation of volatile sulfur compounds in Chardonnay and Shiraz wines post-bottling. *Food Chem.* **2016**, *207*. [CrossRef] [PubMed]

74. Bekker, M.Z.; Wilkes, E.N.; Smith, P.A. Evaluation of putative precursors of key 'reductive' compounds in wines post-bottling. *Food Chem.* **2018**, *245*, 676–686. [CrossRef] [PubMed]

75. Bachhawat, A.K.; Ganguli, D.; Kaur, J.; Kasturia, N.; Thakur, A.; Kaur, H.; Kumar, A.; Yadav, A. Glutathione production in yeast. In *Yeast Biotechnology: Diversity and Applications*; Satyanarayana, T., Kunze, G., Eds.; Springer: Dordrecht, The Netherlands, 2009; Chapter 13; pp. 259–278.

76. Penninckx, M.A. Short review on the role of glutathione in the response of yeasts to nutritonal, environmental, and oxidative stresses. *Enzyme Microb. Technol.* **2000**, *26*, 737–742. [CrossRef]

fermentation

MDPI

Article

Comparison of Different Extraction Methods to Predict Anthocyanin Concentration and Color Characteristics of Red Wines

Stephan Sommer * and Seth D. Cohen

Fermentation Sciences, Appalachian State University, 730 Rivers St., Boone, NC 28608, USA; cohensd@appstate.edu
* Correspondence: sommers@appstate.edu; Tel.: +1-828-262-8136

Received: 10 May 2018; Accepted: 4 June 2018; Published: 7 June 2018

Abstract: Red wines ferment in contact with skins to extract polyphenols and anthocyanins that help build, establish, and stabilize color. Concentration and composition vary among genera, species, and cultivars. For this study, 11 grapes representing *Vitis vinifera* (Cabernet Sauvignon, Merlot, Cabernet Franc, Barbera, Syrah, Petite Sirah, Mourvedre), *Vitis labrusca* (Concord), *Muscadinia rotundifolia* (Noble), and French-American hybrids (Marquette, Chambourcin) were selected. All cultivars were fermented on skins while color extraction was monitored daily. Each grape was also extracted using six different methods (microwave, and ultrasound assisted, Glorie procedure, ITV Standard (Institut Technique de la Vigne et du Vin), AWRI method (Australian Wine and Research Institute), solvent extraction of skins) and compared to color characteristics of the wines produced by fermentation. Results show that the extraction pattern varies among cultivars. Post-fermentation maceration, pressing, and sulfur dioxide addition lead to color loss up to 68 percent of the original maximum with the highest loss for native American grapes and hybrid varieties. Extraction procedures over-estimate color in the finished wine but are more accurate if compared to peak extraction levels during fermentation. Color loss and suitability of different extraction procedures to predict color characteristics of fermented wine strongly depend on the complexity of the anthocyanin spectrum and therefore the cultivar used.

Keywords: anthocyanins; extraction; red wine; color; ultrasound; microwave

---

1. Introduction

Red wines are usually fermented in contact with skins and seeds to extract polyphenols, including anthocyanins, that help to build and stabilize color. Concentration and composition of anthocyanins vary significantly among genera, species, and cultivars [1,2]. The chemical structure of anthocyanins is also very variable and directly influences extractability, solubility, and color characteristics in juice and wine [3,4]. While most cultivars of *Vitis vinifera* have only simple anthocyanin glucosides and acylated derivatives [5], species like *Vitis labrusca* or other genera like *Muscadinia rotundifolia*, or hybrid grapes can display a wider range of anthocyanins including diglycosides of variable composition and structure [2,6]. Extraction kinetics during fermentation are strongly dependent on cultivar, wine style, fermentation conditions, pH and degree of ripeness [4,7], which makes the prediction of color characteristics in the finished wine based on original grape composition very challenging. In addition to that, anthocyanins and other polyphenols in wine start to polymerize, oxidize, and react with other wine components immediately after their extraction [8], which adds an additional challenge to any prediction model. Several polyphenol extraction approaches have been described in the literature that range from solvent based methods [9–11] to combinations between physical treatments with heat, microwaves, or ultrasound with mild solvents [12–14]. Some comparative studies suggest that selected extraction methods can be used to predict the amount of extractable polyphenols [4,15], however, most of these projects only looked at a limited number of *Vitis vinifera* cultivars and only compared finished wines with the predicted polyphenol concentrations. Anthocyanin extraction dynamics during fermentation have been described for *Vitis vinifera* only by Glorie in 1993 [3] and are usually not part of extraction prediction studies. The most frequently used extraction methods in the wine industry are the ITV Standard method [9], the extraction according to Glorie [10], and the method suggested by the AWRI [11]. Previous studies found that, despite the considerable time commitment, Glorie's method is best suited to predict the color characteristics of red wine [4]. However, all methods lead to a significant over-extraction of phenolic substances compared to wine after fermentation, most likely because all extractions are performed with grape paste from a blender, which also leads to a complete destruction and extraction of seeds [4,15]. Microwave-assisted or ultrasound-assisted extraction techniques have been described as an alternative for grapes and other polyphenol-rich material [13,14,16]. Microwave assisted extraction can be viewed as more advanced than traditional solvent extraction methods because the matrix is heated internally and externally without a thermal gradient. Moisture inside and outside the plant cells evaporates, which produces tremendous pressure on the cell structure. When the cell walls rupture, cell material including anthocyanins leaches into the solvent until equilibrium is reached [13]. A similar principle is found with ultrasound-assisted extraction where the treatment increases mass transfer rates by cavitation forces, where bubbles explosively collapse and generate enough force to cause cell rupture [12]. With these physically assisted methods, solvent use and time commitment can be significantly reduced.

The goal of this study was to show color extraction dynamics of 11 grapes during fermentation and compare color characteristics and anthocyanin profiles of the wines to five of the most common extraction methods.

2. Materials and Methods

2.1. Winemaking

2.1.1. Grapes

For this study, 11 grapes representing *Vitis vinifera* (Cabernet Sauvignon, Merlot, Cabernet Franc, Barbera, Syrah, Petite Sirah, Mourvedre), *Vitis labrusca* (Concord), *Muscadinia rotundifolia* (Noble), and French-American hybrids (Marquette, Chambourcin) were selected. Grapes were sourced from North Carolina and California based on availability and condition. Berries were destemmed by hand and only healthy berries were used for the experiments. Half of the destemmed berries was mashed in one-gallon glass jars for on-skin fermentation. The other half was stored as whole berries at −20 °C for different extraction experiments. All fermentations and extractions were performed in duplicates.

2.1.2. Small Scale Fermentations

Hand-crushed berries of each grape variety were inoculated with 20 g/hL *Saccharomyces cerevisiae* yeast (NT 50, Anchor, Johannesburg, South Africa) and fermented at constant temperature of 20 °C for 14 days. Color characteristics of all ferments were checked daily by spectrophotometry (DU 720 UV-Vis spectrophotometer, Beckman Coulter, Brea, CA, USA) at 420, 520, and 620 nm to monitor extraction. Finished wines were pressed and clarified by centrifugation 4000 RPM (3085 RCF) (Allegra X-22, Beckman Coulter, Brea, CA, USA). 100 mg/L sulfur dioxide was added to half the wine volume to assess anthocyanin bleaching effects. These samples were stored at 4 °C until further analysis. The remaining volume was stored without sulfur dioxide at −20 °C.

2.2. Extraction Methods

2.2.1. Solvent Based Extraction of Grape Skins

The solvent based total extraction of anthocyanins from separated grape skins was based on Ageorges et al. (2006) [17] with minor modifications. Frozen berries were peeled to fully separate the skins from pulp and seeds. Fresh skin weight was recorded and half of the skin material was dried in a laboratory oven at 60 °C for three days in aluminum containers until no more weight loss could be observed. Skin dry weight was recorded. The remaining fresh skins were pulverized under liquid nitrogen, and extracted. 100 mg of berry powder was mixed with 400 µL of methanol/water/trifluoroacetic acid (70:30:1, $v/v/v$) as suggested by Barnes, Nguyen, Shen, and Schug (2009) [18] and sonicated for 5 min. The extract was centrifuged at 13,000 RPM (16,060 RCF) at 5 °C for 10 min (Biofuge Fresco, Heraeus Instruments, Hanau, Germany). The extraction was repeated and the two supernatants were combined and stored at −20 °C until further LC analysis.

2.2.2. Microwave Assisted Extraction

The microwave extraction procedure was based on Liazid et al. (2011) [14]. 50 g of berries were put into a kitchen blender (Mini-Prep Processor DLC-1BCH, Cuisinart, East Windsor, NJ, USA) and mashed on a low pulsed level to break pulp and skin but leave the seeds intact. The mash was then quantitatively transferred into a 120 mL plastic tube. 11 mL of 100% ethanol (Koptec, King of Prussia, PA, USA) were added to simulate wine extraction conditions and the tube was closed with a lid. After the microwave (Daewoo Electronics KOR-63D5 9, Seoul, Korea) was set to 700 W output and samples were microwaved for 2.5 min with short interruptions for stirring every full minute and after completed extraction. Juice was then decanted under low manual pressure with a potato ricer and centrifuged at 4200 RPM (3400 RCF) for 10 min. Extracts were transferred into 2 mL centrifuge tubes and centrifuged (Biofuge Fresco, Heraeus Instruments, Hanau, Germany) with 13,000 RPM (16,060 RCF) at 5 °C for 10 min. Spectrophotometric absorbance was analyzed at 280, 420, 520, and 620 nm. An aliquot of each extraction was stored at −20 °C for further LC analysis.

2.2.3. Ultrasound Assisted Extraction

Ultrasound assisted extraction was based on a method by Corrales et al. (2008) [12]. 50 g of berries were put into a kitchen blender (Mini-Prep Processor DLC-1BCH, Cuisinart, East Windsor, NJ, USA) and mashed on a low pulsed level to homogenize pulp and skin but leave the seeds intact. The mash was quantitatively transferred into a 120 mL plastic tube and 11 mL of 100% ethanol (Koptec, King of Prussia, PA, USA) were added to simulate wine extraction conditions. The tubes were placed into the sonicator (Branson 1210, 40 kHz, Danbury, CT, USA) for 60 min while stirring the mash every 10 min. Juice was then decanted under low manual pressure with a potato ricer and centrifuged at 4200 RPM (3400 RCF) for 10 min. Extracts were transferred into 2 mL centrifuge tubes and centrifuged (Biofuge Fresco, Heraeus Instruments, Hanau, Germany) with 13,000 RPM (16,060 RCF) at 5 °C for 10 min. Spectrophotometric absorbance was analyzed at 280, 420, 520, and 620 nm. An aliquot of each extraction was stored at −20 °C for further LC analysis.

2.2.4. ITV Standard Extraction Method

The ITV extraction method was adapted from the description of Cayla et al. (2002) [9]. 50 g of berries were put into a kitchen blender (Mini-Prep Processor DLC-1BCH, Cuisinart, East Windsor, NJ, USA) and mashed on a low pulsed level to homogenize pulp and skin but leave the seeds intact. The mash was then quantitatively transferred into a 120 mL plastic tube and 15 mL of 96% ethanol and 85 mL of 0.1% hydrochloric acid (both VWR International, Radnor, PA, USA) were added. The mixture was incubated for one hour and shacked every 15 min. Juice was then decanted under manual pressure with a potato ricer and centrifuged at 4200 RPM (3400 RCF) for 10 min. The supernatant was decanted and analyzed at 420 and 620 nm. Every extract was then diluted 1:20 with the extraction solution (ethanol and hydrochloric acid). Spectrophotometric absorbance of the diluted extracts was analyzed at 280 and 520 nm. An aliquot of each undiluted extraction was stored at −20 °C for further LC analysis.

2.2.5. AWRI Based Extraction Method

The extraction method for polyphenols used by the AWRI was modified from the description by Iland et al. (2004) [11]. 20 g of berries were put into a kitchen blender (Mini-Prep Processor DLC-1BCH, Cuisinart, East Windsor, NJ, USA) and mashed on a low pulsed level to homogenize pulp and skin but leave the seeds intact. 4 g of the mash were then quantitatively transferred into a 50 mL plastic centrifuge tube. 4.4 mL of a 10 N hydrochloric acid were added to 50% ethanol in water (all VWR International). 20 mL of this solution were then added to the sample and agitated on a shaker table at 30 RPM for one hour at room temperature with more intense manual shacking every 15 min. The samples were centrifuged at 4200 RPM (3400 RCF) for 10 min. Spectrophotometric absorbance of the supernatant was analyzed at 420 and 620 nm with extraction solution as a reagent blank. Each sample was then diluted with 1% hydrochloric acid and absorbance was analyzed at 280 and 520 nm with the 1% HCl as a reagent blank. An aliquot of each undiluted extraction was stored at −20 °C for further LC analysis.

2.2.6. Glorie Extraction Assay

The method for the extraction of anthocyanins and polyphenols developed by Glorie (1984) [10] was adapted with modifications from the description by Kontoudakis et al. (2010) [4]. 100 g of berries were put into a kitchen blender (Mini-Prep Processor DLC-1BCH, Cuisinart, East Windsor, NJ, USA) and mashed on a low pulsed level to homogenize pulp and skin but leave the seeds intact. Two different pH solutions were prepared, a pH 1 solution with 0.3 M oxalic acid (Alfa Aesar, Ward Hill, MA, USA) adjusted with hydrochloric acid and a pH 3.2 with 0.3 M phosphoric acid (VWR International) adjusted with 10 M sodium hydroxide. 50 mL of the pH 1 solution was added to 50 g of grape mash and 50 mL of the pH 3.2 solution was added to the other 50 g of the grape mash. Both were incubated for 4 h at room temperature with shaking every 30 min. Both sample sets were centrifuged at 4200 RPM (3400 RCF) for 10 min. 21 mL extraction solvent (5 mL ethanol and 16.7 mL 12% HCl in 100 mL water) were added to 1 mL of each pH sample supernatant. 4 mL of a 15% (w/w) sulfur dioxide were added to 10 mL of each sample (sulfured samples), while 4 mL water were added to an additional 10 mL of each sample (native samples). All samples were incubated for 15 min and room temperature and analyzed at 280, 420, 520, and 620 nm against the dilution solution (ethanol and HCl in water) as a reagent blank. Extractable and potential anthocyanins were calculated using Formulas (1) and (2). An aliquot of each extraction was stored at −20 °C for further analysis.

$$\text{Potential anthocyanins [mg/L]} = [A_{520} \, (\text{pH 1}) - A_{520} \, (\text{pH 1; SO}_2)] \times 875 \tag{1}$$

$$\text{Extractable anthocyanins [mg/L]} = [A_{520} \, (\text{pH 3.2}) - A_{520} \, (\text{pH 3.2; SO}_2)] \times 875 \tag{2}$$

2.3. Analytical Tools

2.3.1. Photometric Quantification of Polyphenols and Anthocyanins

For quantitative analysis of anthocyanins and polyphenols in wines and extracts, a photometric approach was taken as suggested by Fragoso et al. (2010) [15]. The absorbance at 280 and 520 nm was analyzed for all wines and extracts and compared to a corresponding calibration curve. Samples were diluted with tartrate buffer at pH 3.2 (5 g/L tartaric acid, pH adjusted to 3.2 with 1.0 M NaOH) to fit the linear absorbance range. The polyphenol calibration curve at 280 nm was prepared from a 5.06 g/L gallic acid (gallic acid anhydrous 99%, Alfa Aesar, Heysham, UK) stock solution in 50% ethanol (200 proof, Koptec, King of Prussia, PA, USA) and diluted with tartrate buffer at pH 3.2 into four calibration levels. The anthocyanin calibration curve at 520 nm was prepared from a 2.1 g/L malvidin-3-glucoside (Oenin chloride, Fluka, St. Louis, MO, USA) stock solution in 1% hydrochloric acid (hydrochloric acid 36.5–38%, ACS grade, VWR International) and diluted with tartrate buffer at pH 3.2 into four calibration levels.

2.3.2. Berry, Juice, and Wine Analyses

Standard juice and wine analysis for density, sugar concentration, total acidity, and pH was performed using FT-MIR spectroscopy (FT 2, Foss, Hillerød, Denmark). Nitrogen attributes in the juice were analyzed using spectrophotometry with a commercial kit for ammonium (Ammonia Assay Kit, Megazyme, Bray, Ireland), and the NOPA method published by Dukes and Butzke (1998) for α-amino acids [19]. Total yeast available nitrogen (YAN) was adjusted to 10 mg/L per degree Brix at the second day of fermentation using diammonium phosphate. Total phenolics were analyzed as gallic acid equivalents (Folin−Ciocalteu) as described by Singleton and Rossi (1965) [20]. Total tannin content was determined using a methyl cellulose precipitations assay as described by Sarneckis et al. (2006) [21].

2.3.3. Analysis of Anthocyanins by LC-DAD

The analysis of anthocyanins by LC-DAD at 520 nm was adapted from Rentzsch et al. (2009) [22]. The LC instrument used was an Ultimate 3000 system (Dionex, Sunnyvale, CA, USA) with a Diode Array Multiwavelength Detector. 10 µL of sample was injected onto a Kinetex® 2.6 µm F5 100 Å separation column with the dimensions 50 mm × 2.1 mm (Phenomenex, Torrance, CA, USA). The oven temperature was set to 50 °C. A binary gradient at a constant flow of 0.6 mL/min was used with eluent A being water with 5% acetonitrile and a 10 mM KH_2PO_4/H_3PO_4 buffer and eluent B being a 50:50 acetonitrile–water mixture with a 10 mM KH_2PO_4/H_3PO_4 buffer. Conditions started at 90% eluent A decreased to 85% by 5 min, from 85% to 70% after 12 min, and from 70% to 60% by 15 min, and to 40% after 18 min. After 20 min, it was then set back to initial conditions (90% eluent A) for one minute to reach equilibrium (total analysis time 22 min).

Instrument control and data acquisition were performed with Chromeleon Version 6.80 (Dionex). Anthocyanins were quantified as malvidin-3-glucoside equivalents (standard purchased from Sigma-Aldrich, St. Louis, MO, USA).

2.4. Statistical Analyses

Data handling and statistical analysis were performed using SigmaPlot 12.5 (Systat Software Inc., San Jose, CA, USA) and XLstat 2018.3 (Addinsoft, New York, NY, USA).

3. Results and Discussion

Fermentations finished within 10 days varying with sugar level, which was greatly dependent on growing region and grape cultivar. However, extended macerations continued for a total duration of 14 days to standardize the extraction prior to pressing. Table 1 shows sugar, nitrogen, and acid levels prior to fermentation and prior to diammonium phosphate additions. Some cultivars did not require nitrogen addition, so diammonium phosphate (DAP) was added to deficient trials to reach the recommended 10 mg/L nitrogen per percent sugar.

All wines finished fermentation and fermented to dry. Table 2 shows analytical data collected from wines after pressing and clarification. Since berries were destemmed manually and mashed after separating them into duplicates, the solid-to-liquid ratio was not affected by processing. As a result, the variability among experimental duplicates is minimized and reflects only minor differences and inconsistencies within clusters of one cultivar. Since these fermentations were performed on a four liter scale, inconsistencies within clusters are exaggerated compared to a large scale production where differences are neutralized in a larger volume. The rate of ethanol production and therefore polyphenol extraction depends on the initial sugar concentration but also the solid-to-liquid ratio [23]. Total tannins and total phenols were analyzed as catechin and gallic acid equivalents, respectively. The numbers fall into the range that was previously reported for wines that were made from these cultivars.

Table 1. Juice analysis data prior to fermentation obtained by FT-MIR spectroscopy and enzymatic assays (for nitrogen parameters only).

Cultivar	Density	Sugar [% Bx]	Total Acid [g/L]	pH	NOPA [mg/L]	NH$_4$ [mg/L]	YAN [mg/L]
Mourvedre	1.0838	20.2	5.5	3.58	265	60	325
Cabernet Franc	1.0744	18.1	9.7	3.28	81	14	95
Marquette	1.0799	19.3	14.0	3.19	270	83	353
Syrah	1.1021	24.2	4.3	3.63	122	10	132
Merlot	1.0843	20.3	4.8	3.69	111	37	148
Barbera	1.1119	26.3	6.3	3.72	226	36	262
Petite Sirah	1.1005	23.9	4.5	3.90	164	44	208
Cabernet Sauvignon	1.0934	22.3	5.8	3.74	139	24	163
Chambourcin	1.1038	24.6	5.6	3.77	192	40	232
Noble	1.0665	16.2	8.7	3.04	46	12	58
Concord	1.0645	15.8	2.3	3.36	93	14	107

Table 2. Analytical data of wines after fermentation (standard deviation represents experimental duplicates).

Cultivar	Density	Ethanol [g/L]	Total Acid [g/L]	pH	Total Phenols [mg/L]	Total Tannins [mg/L]
Mourvedre	0.9965 ± 0.0004	83.7 ± 1.1	5.0 ± 0.1	4.0 ± 0.1	758 ± 32	932 ± 36
Cabernet Franc	0.9972 ± 0.0001	75.7 ± 0.4	7.7 ± 0.2	3.4 ± 0.1	823 ± 11	1154 ± 6
Marquette	0.9996 ± 0.0001	76.8 ± 0.1	8.9 ± 0.2	3.5 ± 0.1	519 ± 86	463 ± 158
Syrah	0.9921 ± 0.0001	111.2 ± 2.1	5.5 ± 0.1	3.5 ± 0.1	717 ± 75	976 ± 80
Merlot	0.9950 ± 0.0002	89.6 ± 0.6	5.1 ± 0.1	3.7 ± 0.1	1194 ± 99	1621 ± 131
Barbera	0.9933 ± 0.0001	118.2 ± 1.2	6.3 ± 0.2	3.4 ± 0.1	601 ± 2	332 ± 9
Petit Sirah	0.9934 ± 0.0001	107.1 ± 3.6	5.1 ± 0.1	3.8 ± 0.1	972 ± 135	1224 ± 312
Cabernet Sauvignon	0.9958 ± 0.0001	95.0 ± 0.9	5.8 ± 0.2	4.0 ± 0.1	1071 ± 39	1434 ± 104
Chambourcin	0.9946 ± 0.0003	107.0 ± 1.7	5.8 ± 0.1	3.6 ± 0.1	901 ± 52	942 ± 116
Noble	1.0015 ± 0.0001	68.2 ± 0.3	8.7 ± 0.2	2.9 ± 0.1	1858 ± 417	978 ± 77
Concord	0.9993 ± 0.0002	62.3 ± 0.2	7.8 ± 0.1	2.9 ± 0.1	688 ± 43	780 ± 119

In order to evaluate the levels of total phenols and total tannins after fermentation, it is important to know the relationship between extractable surface area (berry skin) and average berry volume. This information is commonly provided by the 100-berry-weight which leads to the average weight per berry and the weight of the skin in fresh and dried condition. This information is provided in Table 3 for all grape cultivars studied. Because of large differences in overall berry volume, the average weight per berry does not correlate with the fresh skin weight. In fact, berry weight shows much more variability than skin weight. The total amount of extractable anthocyanins and polyphenols shows a similar pattern to the concentrations that were actually analyzed in the wines after fermentation of the skins. However, some variability is caused by a changing solid-to-liquid ratio based on berry volume.

Table 3. Berry and skin weights of all analyzed grapes cultivars including the total extractable anthocyanins and polyphenols obtained by solvent extraction (standard deviation represents experimental duplicates).

Cultivar	100 Berry Weight	Average Weight per Berry	Skin Weight per Berry	Skin Dry Weight per Berry	Extractable Anthocyanins	Extractable Polyphenols
	[g]	[g]	[g]	[g]	[mg/g Fresh Skin]	[mg/g Fresh Skin]
Mourvedre	219.6 ± 17.4	2.20 ± 0.17	0.565 ± 0.025	0.165 ± 0.007	0.37 ± 0.04	1.12 ± 0.35
Merlot	149.5 ± 1.9	1.50 ± 0.02	0.647 ± 0.032	0.211 ± 0.009	0.83 ± 0.09	1.56 ± 0.22
Cabernet Franc	139.4 ± 1.4	1.39 ± 0.01	0.638 ± 0.029	0.196 ± 0.009	1.44 ± 0.18	1.98 ± 0.64
Marquette	105.6 ± 7.8	1.06 ± 0.08	0.591 ± 0.032	0.154 ± 0.007	0.62 ± 0.11	1.23 ± 0.35
Noble	305.0 ± 3.2	3.05 ± 0.03	0.697 ± 0.063	0.251 ± 0.013	2.91 ± 0.41	3.90 ± 0.81
Petite Sirah	107.3 ± 7.6	1.07 ± 0.08	0.660 ± 0.030	0.230 ± 0.012	1.83 ± 0.08	3.41 ± 0.63
Chambourcin	228.2 ± 0.9	2.28 ± 0.01	0.612 ± 0.043	0.219 ± 0.009	1.37 ± 0.21	1.88 ± 0.53
Cabernet Sauvignon	143.3 ± 0.1	1.43 ± 0.01	0.608 ± 0.031	0.205 ± 0.010	1.46 ± 0.17	2.25 ± 0.21
Barbera	149.6 ± 2.0	1.50 ± 0.02	0.685 ± 0.041	0.242 ± 0.012	0.66 ± 0.09	1.03 ± 0.16
Syrah	153.5 ± 3.7	1.53 ± 0.04	0.708 ± 0.042	0.214 ± 0.011	1.62 ± 0.23	2.19 ± 0.42
Concord	300.2 ± 0.4	3.00 ± 0.01	0.858 ± 0.069	0.305 ± 0.015	1.29 ± 0.12	1.85 ± 0.66

Although the amounts of extractable polyphenols and anthocyanins vary significantly among grape species and cultivars, the overall extraction pattern follows the same general trend. Figure 1 shows the extraction of brown, red, and blue color during fermentation and maceration of all studied cultivars. The percentage of brown relative to total color is decreasing after fermentation started and reaches a minimum after three to five days. The slow increase after that until day 14 is probably due to the slow development of polymeric pigments. Red color is constantly extracted in the first two to three days and stays at a stable level until day 14. Blue color remains stable throughout fermentation, with the exception of Noble that shows a decrease in blue color in the beginning of fermentation.

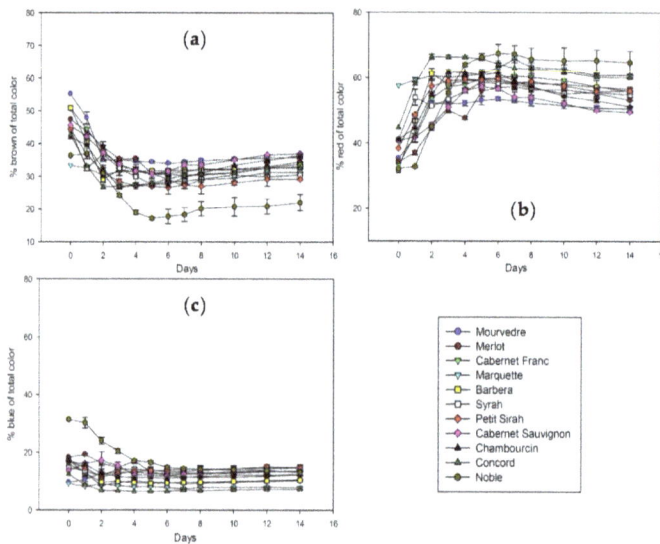

Figure 1. Color extraction in percent of total color during fermentation in contact with the skins analyzed by spectrophotometry at 420 nm (**a**); 520 nm (**b**); and 620 nm (**c**). Absorbance is calculated for 10 mm pathlength.

While the percentage of color does not reveal major differences, the total color shows more variability among grape varieties. Figure 2 summarizes the total color of all wines observed, expressed as the sum of photometric absorbances at 420, 520, and 620 nm. Noble as the only *M. rotundifolia* behaves differently due to a much higher overall extraction level, so Figure 2a shows all the wines except Noble improving the visibility of differences.

It was described before that anthocyanins are not all equally well extractable from berry skin [3,7]. The differences in extractability of anthocyanins show in the total time it takes to reach the color extraction maximum. While some cultivars like Cabernet Franc and Mourvedre need seven days or more to reach that point, color extraction of Concord peaks after only three days. Previous studies associated the extractability with the level of ripeness [4], in this study however, the grape species and cultivar also has an effect. The other main difference among samples is the rate of color loss after reaching that maximum. Although Concord color is extracted very fast, it also shows the fastest rate of loss with more than 40% until the end of maceration on day 14. The reasons for that are speculative at this point but might be related to the anthocyanin spectrum and the presence or absence of stabilizing factors like polysaccharides and other colorless polyphenols. The biggest color losses in the first 14 days of our experiments can be observed in North American *Vitis* species and hybrids. Concord (40.2%), Chambourcin (35.8%), and Marquette (26.5%) seem to have a less stable color than most *Vitis vinifera* grapes which lost an average of 15.1% in the same 14 days. There is of course variability among European grapes as well, a trend however can be hypothesized based on the present data.

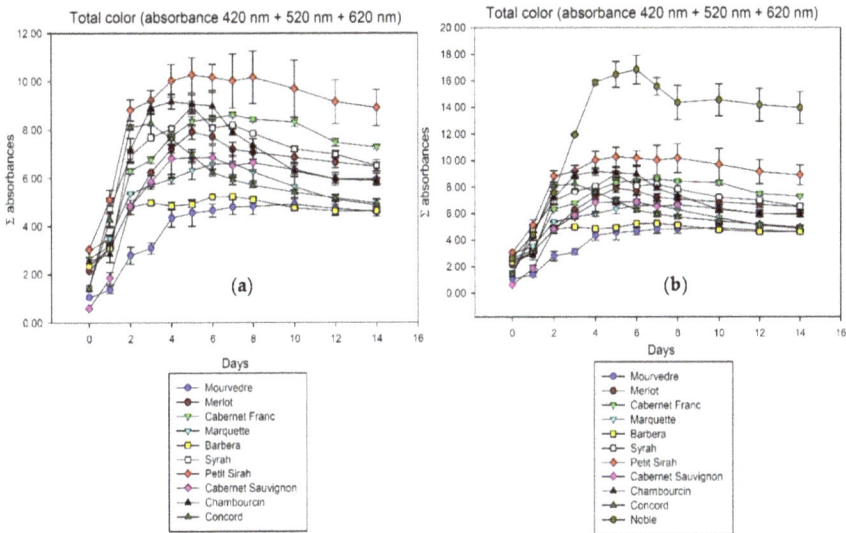

Figure 2. Total color extraction during fermentation in contact with the skins analyzed by spectrophotometry at 420, 520, and 620 nm. Cumulated absorbance is calculated for 10 mm pathlength ((**a**) without Noble; (**b**) including Noble).

After 14 days of fermentation and extended maceration, all grapes were pressed, clarified and stabilized by sulfur dioxide addition. Since any physical force changes the extraction kinetics and SO_2 has a bleaching effect on monomeric anthocyanins [24], the total color change was further monitored throughout the process.

Figure 3 and Table 4 illustrate the main findings. Figure 3 does not include data from the *M. rotundifolia* grape Noble due to differences that qualified the data points as outliers and changed the scaling. The boxplots show brown, red, blue, and total color analyzed at the appropriate wavelengths after fermentation, after pressing, and after sulfur dioxide addition. The color loss is most prominent in the blue spectrum but is very consistent across the whole range. Table 4 shows the individual color losses at each step of the winemaking process, including the days up to the color extraction maximum at each wavelength.

In order to predict the color characteristics and the amount of extractable phenolic material, the most common extraction procedures were performed and compared to the wines after fermentation. The main differences between traditional extraction techniques are pH, time, solvent strength, and temperature. For these experiments, we selected a large variation among these factors with ethanol being the main solvent and pH adjustments ranging from pH 1 to native pH around 3.9. Since extracted anthocyanins and total phenolics were quantified as malvidin-3-glucoside and gallic acid equivalents, respectively, the results can be directly compared to the wines after fermentation. Table 5 shows the extraction results for all grape cultivars.

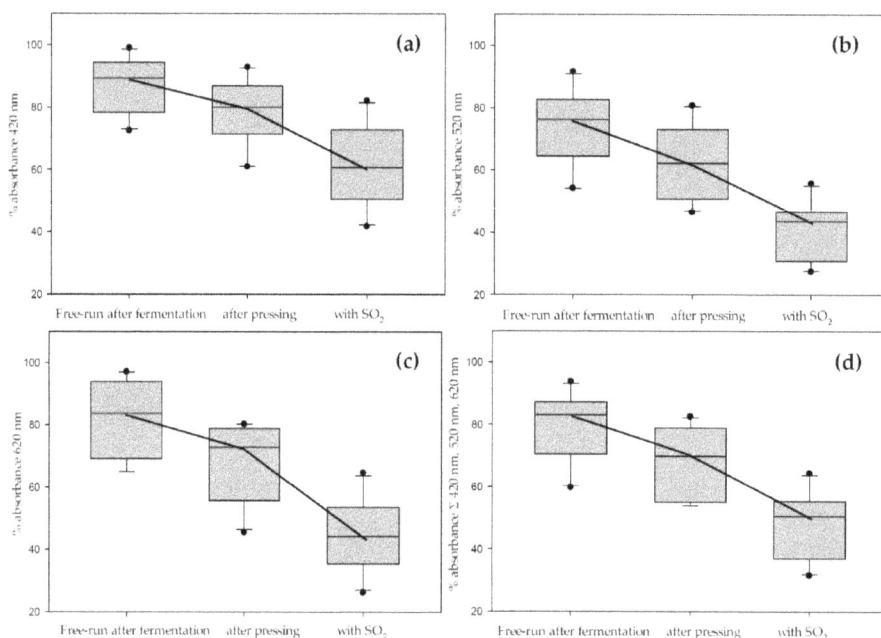

Figure 3. Total color loss during vinification in contact with the skins analyzed by spectrophotometry at 420 nm (**a**); 520 nm (**b**); 620 nm (**c**); and as sum of all three wavelengths (**d**). Cumulated absorbance is calculated for 10 mm pathlength (data does not include Noble).

Table 4. Color characteristics of studied grapes varieties during fermentation and wine production analyzed by UV-Vis spectrophotometry.

Cultivar	Days to Maximum Extraction				Average Color Loss in Free-Run until Pressing [%]			
	420 nm	520 nm	620 nm	Total Σ	420 nm	520 nm	620 nm	Total Σ
Mourvedre	10	7	10	10	4.6	8.4	3.7	6.3
Merlot	4	5	4	5	12.6	22.2	35.0	18.2
Cabernet Franc	7	7	6	7	5.8	22.3	7.0	15.7
Marquette	5	7	6	7	21.7	31.9	25.6	26.5
Barbera	7	6	7	7	9.0	14.8	3.0	11.6
Syrah	2	5	5	5	22.1	32.1	22.7	27.3
Petit Sirah	5	5	8	5	6.2	18.2	8.7	13.3
Cabernet Sauvignon	10	5	3	6	1.0	25.0	10.1	13.1
Chambourcin	5	4	3	4	14.8	45.7	29.5	35.8
Noble	4	6	5	6	−1.9	20.4	32.5	17.1
Concord	3	3	3	3	27.6	45.8	35.0	40.2
Cultivar	Average Color Loss in Press-Wine [%]				Average Color Loss after SO$_2$ and 60 Days [%]			
	420 nm	520 nm	620 nm	Total Σ	420 nm	520 nm	620 nm	Total Σ
Mourvedre	14.2	19.3	20.6	17.5	35.2	54.1	52.0	47.1
Merlot	21.3	36.3	45.4	30.5	37.6	56.4	63.9	49.9
Cabernet Franc	20.7	39.0	39.1	33.3	48.4	68.8	66.6	62.4
Marquette	14.7	39.8	28.4	29.8	28.0	56.3	44.1	45.8
Barbera	19.4	22.7	21.2	21.2	41.1	53.9	47.2	49.0
Syrah	38.5	48.6	54.6	46.1	58.4	72.5	73.8	68.4
Petit Sirah	7.2	28.4	19.8	21.2	18.0	44.2	35.4	35.6
Cabernet Sauvignon	9.9	29.9	24.7	20.8	24.6	51.8	48.0	41.1
Chambourcin	25.4	53.5	44.1	44.6	44.9	66.6	59.7	57.0
Noble	−48.3	26.3	35.0	13.3	0.0	15.4	14.5	11.2
Concord	39.1	51.0	25.8	46.1	52.7	69.9	64.0	64.8

Table 5. Total phenol and anthocyanin concentrations in the grape extracts spectrophotometrically quantified as gallic acid and malvidin-3-glucoside equivalents, respectively (standard deviation represents experimental duplicates).

Cultivar	Finished Wines		Microwave Extraction		Ultrasound Extraction		Glorie Method (pH 3.2)		ITV Standard		AWRI Method	
	Total Phenols	Anthocyanins	Total Phenols	Anthocyanins	Total Phenols	Anthocyanins	Total Phenols	Anthocyanins	Total Phenols	Anthocyanins	Total Phenols	Anthocyanins
	[g/L]	[g/L]	[g/L]	[g/L]	[g/L]	[g/L]	[g/L]	[g/L]	[g/L]	[g/L]	[g/L]	[g/L]
Mourvedre	1.02 ± 0.25	0.21 ± 0.01	2.09 ± 0.14	0.35 ± 0.05	2.09 ± 0.07	0.40 ± 0.02	0.44 ± 0.01	0.24 ± 0.01	0.96 ± 0.08	0.37 ± 0.01	0.59 ± 0.08	0.32 ± 0.01
Merlot	1.23 ± 0.08	0.24 ± 0.01	2.21 ± 0.16	0.39 ± 0.03	1.84 ± 0.04	0.37 ± 0.01	0.42 ± 0.19	0.29 ± 0.01	0.91 ± 0.01	0.38 ± 0.01	0.54 ± 0.09	0.32 ± 0.06
Cabernet Franc	0.98 ± 0.01	0.22 ± 0.01	1.83 ± 0.28	0.37 ± 0.01	1.99 ± 0.04	0.43 ± 0.02	0.44 ± 0.02	0.33 ± 0.02	1.14 ± 0.03	0.44 ± 0.01	0.72 ± 0.06	0.39 ± 0.03
Petite Sirah	1.43 ± 0.30	0.34 ± 0.05	2.24 ± 0.10	0.53 ± 0.02	2.37 ± 0.06	0.54 ± 0.03	0.37 ± 0.02	0.34 ± 0.01	0.78 ± 0.07	0.45 ± 0.03	0.90 ± 0.12	0.54 ± 0.08
Cabernet Sauvignon	1.46 ± 0.18	0.24 ± 0.01	1.96 ± 0.29	0.34 ± 0.04	2.05 ± 0.01	0.38 ± 0.01	0.30 ± 0.06	0.25 ± 0.01	0.69 ± 0.01	0.32 ± 0.02	0.56 ± 0.01	0.35 ± 0.01
Barbera	0.75 ± 0.02	0.21 ± 0.01	1.46 ± 0.19	0.33 ± 0.03	1.47 ± 0.10	0.36 ± 0.02	0.22 ± 0.02	0.23 ± 0.01	0.80 ± 0.07	0.31 ± 0.01	0.58 ± 0.01	0.39 ± 0.01
Syrah	1.03 ± 0.29	0.23 ± 0.01	1.78 ± 0.15	0.37 ± 0.04	2.44 ± 0.02	0.49 ± 0.01	0.25 ± 0.14	0.25 ± 0.01	0.80 ± 0.02	0.34 ± 0.02	0.65 ± 0.14	0.38 ± 0.11
Chambourcin	1.64 ± 0.04	0.24 ± 0.01	2.46 ± 0.04	0.39 ± 0.01	2.19 ± 0.22	0.36 ± 0.01	0.59 ± 0.01	0.56 ± 0.05	1.35 ± 0.01	0.70 ± 0.01	0.59 ± 0.16	0.49 ± 0.08
Marquette	1.04 ± 0.08	0.20 ± 0.01	1.86 ± 0.06	0.36 ± 0.01	1.98 ± 0.08	0.35 ± 0.01	0.44 ± 0.01	0.40 ± 0.02	1.07 ± 0.08	0.55 ± 0.01	0.49 ± 0.03	0.39 ± 0.01
Noble	4.87 ± 0.16	0.35 ± 0.01	3.93 ± 0.31	0.30 ± 0.03	3.95 ± 0.23	0.35 ± 0.01	1.26 ± 0.18	1.13 ± 0.12	2.92 ± 0.18	1.61 ± 0.09	0.89 ± 0.28	0.71 ± 0.13
Concord	1.45 ± 0.22	0.19 ± 0.01	2.61 ± 0.13	0.42 ± 0.05	2.32 ± 0.01	0.37 ± 0.01	0.34 ± 0.01	0.23 ± 0.01	1.03 ± 0.04	0.42 ± 0.01	0.41 ± 0.02	0.26 ± 0.02

Most extraction methods used here show the tendency to over-extract anthocyanins while underrepresenting total polyphenols. In order to see correlations and trends, statistical analyses, shown in Table 6, used correlation coefficients and calculated if the observation is significantly similar. The results indicate that most extraction methods correlate significantly with the phenolic composition of the finished wines. The color extraction discussed above however makes it a relatively weak correlation. While anthocyanins and other polyphenols in wine start to polymerize, oxidize, and react with other wine components immediately after their extraction, the extracts do not have the time to mimic that before they are analyzed. Table 6 therefore includes the mentioned color extraction maximum as well, which shows an improved correlation. In fact, when the extracts are compared to the peak color intensity, some of the correlations are highly significant with correlation coefficients above 85%. This observation illustrates the difficulties of predicting color characteristics in real wine fermentations. The factor time and the reactivity of wine polyphenols cannot be factored into the prediction model because it varies with grape cultivar, production method, and the overall redox potential throughout the process.

The microwave and ultrasound assisted extraction methods show no significant correlation with color characteristics but are highly significant for total phenolics. The explanation for this observation is most likely the pretty severe change in extraction conditions compared to wine. Both methods use physical force to break cell walls and could lead to a change in extraction kinetics. Especially for acylated anthocyanins, which were shown be extracted easier than mono-glucosides due to their higher solubility in water [4], a destruction of cell wall material favors their extraction. In addition, acylated anthocyanins are more entrapped in the matrix or form hydrogen bonds with polysaccharides, which can inhibit their extraction [12]. Microwave and ultrasound assisted extractions have the potential to release more phenolic material than solvent based methods by disrupting these structures.

Table 6. Correlation coefficients (R^2) between phenolic characteristics of wines during and after fermentation, and the predicted levels produced by grape extraction (n.s.: not significant, *: $\alpha = 0.05$, **: $\alpha = 0.01$, ***: $\alpha = 0.001$).

Observation	Microwave Assisted Extraction	Ultrasound Assisted Extraction	ITV Standard	AWRI Extraction	Glorie Extraction	Potential Anthocynins (Glorie)	Extractable Anthocyanins (Glorie)
Anthocyanins finished wines	0.199 (n.s.)	0.345 (n.s.)	**0.638 (*)**	**0.797 (***)**	**0.650 (*)**	**0.674 (**)**	**0.650 (*)**
Color maximum during fermentation	0.086 (n.s.)	0.005 (n.s.)	**0.895 (***)**	**0.741 (***)**	**0.870 (***)**	**0.898 (***)**	**0.868 (***)**
Total phenolics finished wines	**0.902 (***)**	**0.924 (***)**	**0.933 (***)**	**0.452 (*)**	**0.909 (***)**		

Numbers in bold indicate statistical significance.

The overall differences in extraction between the methods are best visualized in an overlay of LC chromatograms, shown in Figure 4. For this figure, Chambourcin was chosen because, being a hybrid, the grape has one of the broadest anthocyanin spectrums in this study. Although not all compounds are baseline separated with this quick method, it is obvious that anthocyanins in the second half of the chromatogram are not or only poorly extracted by some of the methods used here. The method that stands out is the AWRI extraction. It was originally not designed to predict anthocyanins and although it shows a good correlation with the red wines after fermentation, it does not nearly extract as much anthocyanins as other methods. The spectrum however and the ratio between single anthocyanins is close to the corresponding wine. Ultrasound and microwave assisted extraction on the other hand severely over-extract color compounds as shown in Figure 4. The ratio is not close to the real wine, which explains the non-existing correlation shown in Table 6.

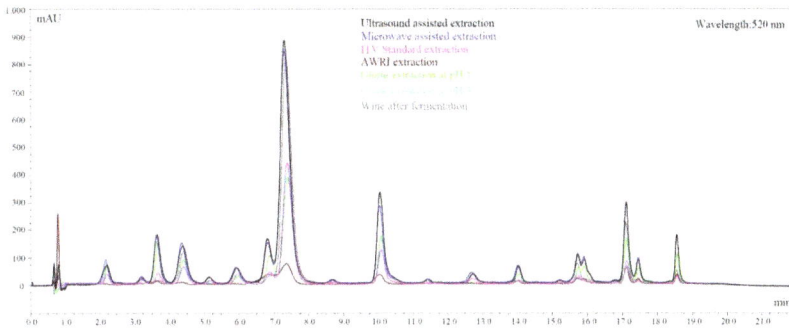

Figure 4. Chromatograms comparing extraction through fermentation and other techniques in Chambourcin grapes.

It can still be used to have a complete spectrum of anthocyanins but is not useful for the prediction of color characteristics. The grape crushing method suggested in our study should be considered, since it was less destructive than other methods. In previous studies, the grape material is often completely destroyed including seed material, which increases the rate of polyphenol extraction and could be less representative of an actual wine fermentation. Figure 5 summarizes these observations in a Principle Component Analysis. Most extraction methods are positively correlated, which implies that the quality of their prediction is similar. The vectors of anthocyanins extracted by microwave and ultrasound are located at a 90 degree angle to the other extractions, showing that there is no correlation between them.

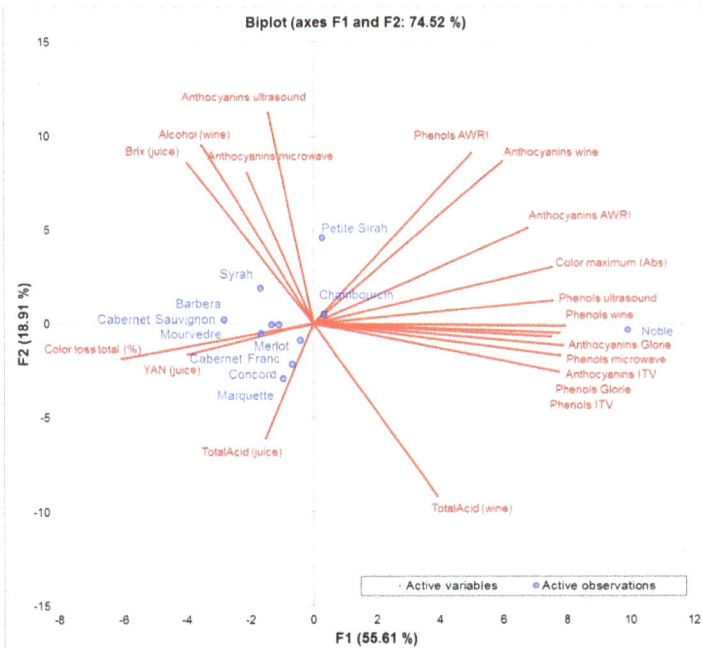

Figure 5. Principle component analysis of all grape cultivars included in this study (observations) and the most important influencing factors and analytical data-points (variables).

4. Conclusions

The overall goal from a winemaker's perspective when evaluating red grapes before harvesting is to predict the phenolic profile and color characteristics of the finished product and determine the harvest date accordingly. This study aimed to provide insight into different laboratory-based extraction methods and compare the extracts to wines that were fermented on the skins. Most previous publications used a limited number of extraction methods and looked primarily at *Vitis vinifera* grapes while we included hybrids and North American grape cultivars as well. During fermentation, all cultivars reached a color extraction maximum after a few days and then started to lose color. This presents the main challenge with predicting color characteristics, because most methods show much better correlations with the wine when they are compared to the maximum color intensity and not the finished wine after processing. The level of anthocyanin extraction as well as color loss due to oxidation, polymerization, pressing, and sulfur addition depends, among previously reported characteristics like the level of ripeness and the grape cultivar. Varieties with a more complex anthocyanin spectrum display a different extraction pattern that cannot be mimicked accurately with some of the extraction methods. Because most methods tend to over-extract phenolic material, the correlation with the peak of color intensity is much better.

Despite the large time commitment, the extraction suggested by Glorie is still the best way to predict extractable anthocyanins, mostly because the method compares different pH conditions in one assay. The ITV Standard method also correlated well with the color extraction maximum but had a larger discrepancy with the wines after processing. Physically invasive methods like ultrasound or microwave assisted extractions are great for extracting more anthocyanins but over-extract to an extent where the spectrum is too different from the corresponding wine to be useful for a prediction model. The correlation with total phenols on the other hand shows promising results. Ultrasound assisted extraction, for example, does not use harsh solvent conditions and consumes very little time and energy, which potentially makes it a very convenient method to predict total extractable phenolics as a cost-effective alternative for smaller wineries.

Author Contributions: Conceptualization, Methodology, Formal Analysis, S.S.; Formal Analysis and Resources, S.D.C.

Acknowledgments: This project was supported through generous grape donations by wineries in North Carolina. We would like to thank Grandfather Vineyards and Winery, Jones von Drehle Vineyards and Winery, Windsor Run Cellars, as well as Ivory Tower Inc. for their generosity and continuous research support.

Conflicts of Interest: The authors declare no conflict of interest.

References

1. Revilla, E.; García-Beneytez, E.; Cabello, F.; Martín-Ortega, G.; Ryan, J.-M.A. Value of high-performance liquid chromatographic analysis of anthocyanins in the differentiation of red grape cultivars and red wines made from them. *J. Chromatogr. A* **2001**, *915*, 53–60. [CrossRef]
2. Wu, X.; Prior, R.L. Systematic identification and characterization of anthocyanins by hplc-esi-ms/ms in common foods in the United States: Fruits and berries. *J. Agric. Food Chem.* **2005**, *53*, 2589–2599. [CrossRef] [PubMed]
3. Glories, Y.; Augustin, M. *Maturité Phénolique du Raisin, Conséquences Technologiques: Application Aux Millésimes 1991 et 1992*; CR Colloque Journée Techn; CIVB: Bordeaux, France, 1993; pp. 56–61.
4. Kontoudakis, N.; Esteruelas, M.; Fort, F.; Canals, J.M.; Zamora, F. Comparison of methods for estimating phenolic maturity in grapes: Correlation between predicted and obtained parameters. *Anal. Chim. Acta* **2010**, *660*, 127–133. [CrossRef] [PubMed]
5. García-Beneytez, E.; Revilla, E.; Cabello, F. Anthocyanin pattern of several red grape cultivars and wines made from them. *Eur. Food Res. Technol.* **2002**, *215*, 32–37. [CrossRef]
6. Huang, Z.; Wang, B.; Williams, P.; Pace, R.D. Identification of anthocyanins in muscadine grapes with hplc-esi-ms. *LWT Food Sci. Technol.* **2009**, *42*, 819–824. [CrossRef]
7. Ribéreau-Gayon, P.; Dubourdieu, D.; Donèche, B.; Lonvaud, A. *Handbook of Enology*; John Wiley & Sons: Hoboken, NJ, USA, 2006; Volume 1.

8. Romero-Cascales, I.; Fernández-Fernández, J.I.; López-Roca, J.M.; Gómez-Plaza, E. The maceration process during winemaking extraction of anthocyanins from grape skins into wine. *Eur. Food Res. Technol.* **2005**, *221*, 163–167. [CrossRef]
9. Cayla, L.; Cottereau, P.; Renard, R. Estimation de la maturité phénolique des raisins rouges par la méthode i.T.V. Standard. *Rev. Fr. d'OEnol.* **2002**, *193*, 10–16.
10. Glories, Y. La couler des vins rouges: 2a. Partie mesure, origine et interpretation. *Connaiss. Vigne Vin* **1984**, *18*, 253–271.
11. Iland, P.; Bruer, N.; Edwards, G.; Weeks, S.; Wilkes, E. *Chemical Analysis of Grapes and Wine: Techniques and Concepts*; Patrick Iland Wine Promotions PTY Ltd.: Campbelltown, Australia, 2004.
12. Corrales, M.; Toepfl, S.; Butz, P.; Knorr, D.; Tauscher, B. Extraction of anthocyanins from grape by-products assisted by ultrasonics, high hydrostatic pressure or pulsed electric fields: A comparison. *Innov. Food Sci. Emerg. Technol.* **2008**, *9*, 85–91. [CrossRef]
13. Li, Y.; Skouroumounis, G.K.; Elsey, G.M.; Taylor, D.K. Microwave-assistance provides very rapid and efficient extraction of grape seed polyphenols. *Food Chem.* **2011**, *129*, 570–576. [CrossRef]
14. Liazid, A.; Guerrero, R.F.; Cantos, E.; Palma, M.; Barroso, C.G. Microwave assisted extraction of anthocyanins from grape skins. *Food Chem.* **2011**, *124*, 1238–1243. [CrossRef]
15. Fragoso, S.; Mestres, M.; Busto, O.; Guasch, J. Comparison of three extraction methods used to evaluate phenolic ripening in red grapes. *J. Agric. Food Chem.* **2010**, *58*, 4071–4076. [CrossRef] [PubMed]
16. Corrales, M.; García, A.F.; Butz, P.; Tauscher, B. Extraction of anthocyanins from grape skins assisted by high hydrostatic pressure. *J. Food Eng.* **2009**, *90*, 415–421. [CrossRef]
17. Ageorges, A.; Fernandez, L.; Vialet, S.; Merdinoglu, D.; Terrier, N.; Romieu, C. Four specific isogenes of the anthocyanin metabolic pathway are systematically co-expressed with the red colour of grape berries. *Plant Sci.* **2006**, *170*, 372–383. [CrossRef]
18. Barnes, J.S.; Nguyen, H.P.; Shen, S.; Schug, K.A. General method for extraction of blueberry anthocyanins and identification using high performance liquid chromatography-electrospray ionization-ion trap-time of flight-mass spectrometry. *J. Chromatogr. A* **2009**, *1216*, 4728–4735. [CrossRef] [PubMed]
19. Dukes, B.C.; Butzke, C.E. Rapid determination of primary amino acids in grape juice using an *o*-phthaldialdehyde/*N*-acetyl-L-cysteine spectrophotometric assay. *Am. J. Enol. Vitic.* **1998**, *49*, 125–134.
20. Singleton, V.L.; Rossi, J.A. Colorimetry of total phenolics with phosphomolybdic-phosphotungstic acid reagents. *Am. J. Enol. Vitic.* **1965**, *16*, 144–158.
21. Sarneckis, C.J.; Dambergs, R.G.; Jones, P.; Mercurio, M.; Herderich, M.J.; Smith, P.A. Quantification of condensed tannins by precipitation with methyl cellulose: Development and validation of an optimised tool for grape and wine analysis. *Aust. J. Grape Wine Res.* **2006**, *12*, 39–49. [CrossRef]
22. Rentzsch, M.; Weber, F.; Durner, D.; Fischer, U.; Winterhalter, P. Variation of pyranoanthocyanins in red wines of different varieties and vintages and the impact of pinotin A addition on their color parameters. *Eur. Food Res. Technol.* **2009**, *229*, 689–696. [CrossRef]
23. Sacchi, K.L.; Bisson, L.F.; Adams, D.O. A review of the effect of winemaking techniques on phenolic extraction in red wines. *Am. J. Enol. Vitic.* **2005**, *56*, 197–206.
24. Jurd, L. Reactions involved in sulfite bleaching of anthocyanins. *J. Food Sci.* **1964**, *29*, 16–19. [CrossRef]

fermentation

MDPI

Article

Microwave-Assisted Extraction Applied to Merlot Grapes with Contrasting Maturity Levels: Effects on Phenolic Chemistry and Wine Color

L. Federico Casassa [1,*], Santiago E. Sari [2], Esteban A. Bolcato [2] and Martin L. Fanzone [3]

[1]   Wine & Viticulture Department, California Polytechnic State University, San Luis Obispo, CA 93407, USA
[2]   Centro de Estudios de Enología, Estación Experimental Agropecuaria Mendoza, Instituto Nacional de Tecnología Agropecuaria (INTA), San Martín 3853, Luján de Cuyo, Mendoza 5507, Argentina; sari.santiago@inta.gob.ar (S.E.S.); bolcato.esteban@inta.gob.ar (E.A.B.)
[3]   Laboratorio de Aromas y Sustancias Naturales, Instituto Nacional de Tecnología Agropecuaria (INTA), San Martín 3853, Luján de Cuyo, Mendoza 5507, Argentina; fanzone.martin@inta.gob.ar
*   Correspondence: lcasassa@calpoly.edu; Tel.: +805-756-2751

Received: 4 January 2019; Accepted: 24 January 2019; Published: 28 January 2019

Abstract: Merlot grapes were harvested with three maturity levels (21.1, 23.1, and 25.1 Brix), and processed with or without the application of microwave-assisted extraction (MW). The detailed phenolic composition and color were followed during winemaking. The MW treatment did not affect the basic chemical composition of the wines. Upon crushing, MW caused a 211% improvement in anthocyanins in the wines of the first harvest and an 89% improvement in the wines of the third harvest. At bottling, MW favored the formation of pyranoanthocyanins and tannin-anthocyanin dimers. Tannin extraction was not affected by MW just after application of this process, but improvements of 30, 20, and 10% on MW-treated wines of the first, second, and third harvest, respectively, were recorded at pressing. The formation of polymeric pigments during aging generally increased along with harvest date and was only favored in MW-treated wines of the first and third harvest, with preferential formation of small polymeric pigments, in accordance with enhanced anthocyanin extraction in these wines. Initial improvements of wine color upon application of MW in the wines of the first, second, and third harvest were of 275, 300, and 175%, respectively. Although these differences subsided or disappeared for the wines of the second and third harvest during aging, the wines of the first harvest treated with MW retained 52% more color than Control wines at day 150 post-crushing. Results suggest the MW treatment was more efficient in extracting and retaining phenolics and color when applied to unripe fruit.

Keywords: anthocyanins; tannins; polymeric pigments; wine color; grape maturity; microwave-assisted extraction; Merlot

1. Introduction

In *Vitis vinifera* L. grapes and their wines, grape maturity plays a key role in defining the phenolic and aromatic composition of the resulting wines [1–3]. Phenolic compounds are secondary metabolites that define critical aspects of red wine sensory composition, including color, taste, and mouthfeel sensations [4]. Phenolic compounds are non-volatile molecules synthetized through the phenylpropanoid pathway [5]. This pathway has its inception shortly after bloom [6], and continues during berry ripening, producing, among other molecules, anthocyanins, compounds responsible for wine color; tannins, compounds responsible for wine textural properties, such as astringency; and flavan-3-ols, compounds responsible for eliciting bitterness sensations in red wines [5,7]. As the relative proportion of anthocyanins (generally confined to the skins of the majority of *Vitis vinifera* L.

varieties), and tannins (located both in skins and seeds) change during ripening, winemakers usually define harvest decisions based on a complex combination of factors that include weather forecast and disease pressure, fruit basic chemistry (including Brix, pH, and titratable acidity), and phenolic composition (including anthocyanins and tannins).

Studies in Cabernet Sauvignon, Merlot, and Tempranillo have found a direct link between grape maturity, usually defined in terms of color and phenolic composition, and wine composition [1,3,8]. This accumulated knowledge has, in turn, allowed the production of wines with specific sensory features based on harvest time. Bindon et al. (2014) [9] studied the effect of five different harvest dates, with sugar levels spanning from 20 to 26 Brix, and corresponding alcohol contents in the finished wines spanning from 12 to 15.5% v/v. It was found that wine anthocyanins and tannins increased with harvest date, as well as an enhanced extraction of skin-derived phenolics [9]. From a sensory standpoint, later harvest dates produced wines with less green or vegetative attributes and more dark fruit attributes [1].

Because grape maturity plays such a critical role in wine composition, it is expected that the outcomes of a given winemaking technique may be different based on when the fruit is harvested. In turn, variations in ethanol due to changes in fruit maturity may have an effect in phenolic extraction as well. For example, a study based on Cabernet Sauvignon and Tempranillo grapes harvested at three different maturity levels and processed with four different maceration lengths (1, 2, 3, and 4 weeks), reported that tannin extraction in Cabernet Sauvignon was enhanced in riper fruit and more so when submitted to longer maceration times. However, fruit ripeness did not affect tannin extraction in Tempranillo wines [8]. Contrastingly, Casassa et al. (2013) [3] investigated the effect of grape maturity and extended maceration (30 days) in Merlot grapes and their corresponding wines, and alcohol contents produced from fruit with two contrasting maturity levels, namely 20.3 and 24.9 Brix, were also compensated during maceration via chaptalization and water addition. It was found that riper fruit produced wines that were fruitier and with fuller mouthfeel, whereas chaptalization of unripe fruit shifted the sensory profile towards that of riper fruit. It was also found that the outcomes of extended maceration, particularly in regards to tannin extraction, were independent of fruit maturity [3]. These discrepancies suggest that variety, fruit maturity and the specific winemaking technique employed all affect the phenolic composition, color, and sensory properties in the finished wines.

Among the bevy of innovative winemaking techniques introduced in the wine industry since the early 90's, thermomaceration (also known as thermovinification) and flash-expansion (also known as flash-détente), are based on heating the crushed grapes before fermentation to favor phenolic extraction. In effect, if phenolic extraction into wine is accepted as a combination of both dissolutive and diffusive processes, application of heat to freshly crushed grapes may boost mass-transfer processes, thus, favoring diffusion in the case of anthocyanins [10]. In addition, temperature may also increase the permeability of cell wall membranes, thereby favoring dissolution [11].

A series of studies using flash-détente applied to Grenache and Carignan grapes, whereby the heated must (95 °C) is released into a vacuum causing cell disorganization, found that this technique increased the extraction of polysaccharides [12] as well as wine phenolics by a factor of 1.2 to 2 [13]. Another study compared the effects of thermomaceration at 60 °C applied to País (also known as Mission) and Lachryma Christi grapes to that of the use of pectolytic enzymes, commonly used as extraction aids in winemaking. Expectedly, thermomaceration favored phenolic extraction and increased antioxidant capacity in both varieties relative to the enzyme treatment, which was attributed to an increased diffusion coefficient [14].

Given the previous findings, winemakers typically restrict the use of thermomaceration or flash-détente to lesser quality grapes that may be, for instance, unripe (thereby lacking phenolic maturity), or affected by fungal diseases. The reasoning behind is that the extractability of phenolic compounds of sensory relevance is lower in unripe fruit or compromised versus fully ripe fruit, and as such, positive improvements in the finished wines could be more effectively achieved by thermally treating unripe grapes.

One problem associated with the use and application of thermomaceration or flash-détente, however, is that both techniques are onerous, requiring steam generators (i.e., large input of water), heat exchangers, and dedicated pumps installed on-site. Conversely, microwave-assisted extraction (MW) is a novel technique that has been recently studied at experimental scale [15]. Industrial-scale MW has been in operation in the food industry over a period of several decades now, but its use in winemaking it is innovative and appears attractive. During MW, the dipolar nature of water molecules caused them to align with the direction of an oscillating electric field created by MW. This alignment occurs at a rate of million times per second and causes internal friction of molecules resulting in volumetric heating [16]. Importantly, MW requires no water input. A preliminary study of the application of MW to Cabernet Sauvignon, Merlot, and Syrah grapes was conducted to explore a potential differential effect of MW on different grapes varieties. This study found an interaction between variety and the MW treatment in that the MW treatment failed to increase color in Cabernet Sauvignon and Merlot, but it did so in Syrah [17]. Another study on Pinot noir, a variety particularly deficient in skin-derived phenolics, reported that MW produced a four-fold increase in tannin concentration, additionally causing a decrease in the native grape yeast-derived populations [18].

In the present work, we examined the effect of MW applied to three batches of Merlot fruit from the same vineyard block harvested at three maturity levels. This included unripe fruit (to simulate fruit harvested earlier due to disease pressure), properly ripe fruit (to simulate production of regular table red wine), and overripe fruit (more representative of current standard winemaking conditions). At each harvest point, triplicate fermentations of untreated musts (i.e., Control wines), and MW-treated musts were followed in terms of their color properties as well as specific phenolic composition throughout winemaking. Herein we present a detailed report of the effect of harvest maturity, MW-assisted extraction and their combination thereof on the chromatic and phenolic composition of Merlot wines. This study aimed to establish the most suitable grape maturity level that may optimize the application of MW for winemaking purposes.

2. Materials and Methods

2.1. Grapes

Own-rooted Merlot grapes (*Vitis vinifera* L.) were obtained from a commercial vineyard located in Luján de Cuyo, Mendoza, Argentina (33° 00′ S, 68° 51′ W). The fruit was sequentially harvested targeting 21, 23, and 25 Brix on February 3, February 27, and March 29, 2015, respectively. At each harvest, a total of 150 kg of fruit were manually harvested to 18-kg plastic boxes for a total of 360 kg. For the grape basic analysis, four independent samples, each of 30 berries, were taken at each harvest analyzed for basic fruit chemistry, including Brix (Atago, Tokyo, Japan), pH (Orion model 701-A, Thermo Scientific, Waltham, MA, USA), and titratable acidity, which was determined manually by titration with 0.1 N NaOH and expressed as g/L of tartaric acid [19] (Table 1).

2.2. Winemaking and Application of Microwave-Assisted Extraction

Upon each harvest, grapes were transported to the experimental winery of INTA. Upon reception, grapes were first crushed and destemmed (Metal Liniers model MTL 12, Mendoza, Argentina), and the musts placed into 20-L food-grade plastic fermentors. The experimental design consisted of two winemaking treatments applied to the fruit of each harvest, namely Control (not microwaved) wines, and microwaved-assisted extraction (MW) wines, with each winemaking treatment replicated three times ($n = 3$), and the MW treatment applied inmediately after crushing. Control wines were produced with a standard SO_2 addition of 50 mg/L, and a maceration length of 14 days at 24.5 ± 2.5 °C; cap management consisted of two daily punch downs (morning and afternoon, 1 min each). For the MW treatments, a commercially available household microwave oven was used. Musts were microwaved at 1200 Watts for 10 min (~ 400 Watts/kg), reaching an average temperature of approximately 40 °C. This working temperature was attained in the whole mass and was calculated in each batch as the average of three independent temperature measurements throughout the mass of must being microwaved (~ 3 kg) (exact average temperature attained $\pm$ standard deviation: $40.16 \pm 3.81\ °C$, $n = 270$). Five batches of MW musts were added per each fermentor for a total of ~ 15 kg of musts/fermentor. Upon cooling, MW wines were added with 50 mg/L of SO_2 and produced with a maceration length of 14 days, following the same cap management regime applied to Control wines. All the fermentors were inoculated approximately 8 h after crushing with a commercial yeast (EC-1118; Lallemand Inc., Copenhagen, Denmark) at a rate of 0.3 g/L. After maceration was completed, free run wines were collected into 20 L glass carboys fitted with airlocks. Malolactic fermentation (MLF) was induced with a commercial *Oenococcus oeni* culture (VP-41, Lallemand Inc., Copenhagen, Denmark). Malic acid composition was followed monthly, and MLF was stopped when there was no significant consumption of malic acid in the wines within a 30-day period. After this, the wines were racked off the lees, adjusted to 30 mg/L of free SO_2, and stored at $1\ °C$ for 45 days to allow tartaric stabilization. After this period, the wines were racked and brought to room temperature for 48 h. Before bottling, free SO_2 was adjusted to 0.5 mg/L of molecular SO_2. The bottles were then stored horizontally in a cellar ($12 \pm 1\ °C$) until needed.

2.3. Wine Basic Analysis

Alcohol content, residual sugars, glucose + fructose, tartaric, citric, lactic and acetic acid, titratable acidity, and glycerol content were measured using a FOSS Wine-Scan (FT120) rapid-scanning infrared Fourier-transform spectrometer (FOSS, Hillerod, Denmark). Malic acid was determined enzymatically (Vintessential Laboratories, Victoria, Australia). pH was measured with an Orion model 701-A (Thermo Scientific, Waltham, MA, USA). Titratable acidity was determined manually by titration with 0.1 N NaOH and expressed as g/L of tartaric acid [19].

2.4. Spectrophotometric Analysis of the Wines

Must and wine samples were filtered through 0.22-mm filters (Fisher Scientific, Fair Lawn, NJ, USA) before analysis. Spectrophotometric measurements were carried out with a Perkin-Elmer Lambda 25 UV-visible spectrophotometer (PerkinElmer, Hartford, CT, USA). Wine color was calculated as the summation of the absorbances at 420, 520, and 620 nm and hue as the ratio of absorbances at 420 and 520 nm [20]. Tannins in the wines were analyzed by protein precipitation [21]. Anthocyanins, small polymeric pigments (SPP), large polymeric pigments (LPP), and total polymeric pigments (SPP + LPP) were measured as previously described [21].

2.5. HPLC-DAD-Mass Spectrometry (MS) Analysis of the Wines

The detailed pigment composition of each of the 18 wines was measured at bottling. Identification and MS confirmation of anthocyanins were performed by HPLC-DAD coupled with electrospray ionization (ESI)-MS as described [22]. The chromatographic system employed was a Perkin-Elmer Series 200 HPLC-DAD equipped with a quaternary pump and an autosampler (PerkinElmer, Hartford, CT, USA). Two mL of wine was filtered through a 0.45-µm pore size nylon membrane (Fisher Scientific, Fair Lawn, NJ, USA), and a 100-µL aliquot was injected onto the column. Separation was performed on a reversed-phase Chromolith Performance C18 column (100 mm × 4.6 mm i.d., 2 µm; Merck (Darmstadt, Germany)) protected with a Chromolith guard cartridge (10 mm × 4.6 mm) at 25 °C. A gradient consisting of solvent A (water/formic acid, 90:10, v/v) and solvent B (acetonitrile) was applied at a flow rate of 1.1 mL/min from 0 to 22 min and 1.5 mL/min from 22 to 35 min as follows: 96 to 85% A from 0 to 12 min, 85 to 85% A from 12 to 22 min, 85 to 70% A from 22 to 35 min; followed by a final wash with 100% methanol and re-equilibration of the column. DAD was performed from 210 to 600 nm, and the quantification was carried out using peak area measurements at 520 nm. In both years, monomeric anthocyanins and anthocyanin derivatives were quantified using malvidin-3-O-glucoside chloride as the standard (Extrasynthèse, Genay, France) and a standard calibration curve for each year ($R^2 = 0.98$ and 0.99 for 2014 and 2015, respectively).

2.6. Data Analysis

The basic composition of the fruit at harvest was analyzed by a one-way analysis of variance (ANOVA). The phenolic, anthocyanin, and chromatic compositions of the wines were analyzed by a one-way ANOVA. When appropriate, a two-way ANOVA separating the effects of the grape maturity and winemaking treatments, and their interaction, was also performed. Fisher's least significant difference (LSD) test was used as a post-hoc comparison of means in all cases with a 5% level for rejection of the null hypothesis. The detailed phenolic, anthocyanin, and pigment composition was further analyzed by simultaneous Student's *t*-tests comparing Control and MW wines of each harvest separately, establishing a 5% level for rejection of the null hypothesis. Data were analyzed with XLSTAT v. 2015 (Addinsoft, Paris, France).

3. Results and Discussion

3.1. Basic Chemical Composition of the Grapes

Merlot fruit was harvested in three separate events to capture a wide range of fruit chemistry based on harvest time. In turn, these chemical changes at the fruit level were expected to impact the outcome of MW-assisted extraction on their respective wines. Overall, the three harvests spanned almost two months, from early February to late March. There were 24 days from the first to the second harvest, and 30 days between the second and third harvest (Table 1). The first harvest yielded fruit considered unripe for standard winemaking conditions for red wine production. The second harvest produced fruit considered ripe whereas the third harvest, while slightly overripe, was also considered fully ripe for current winemaking standards. As intended, as maturity progressed from the first to the third harvest, Brix and pH increased whereas titratable acidity decreased.

Table 1. Harvest date and basic chemical composition of the fruit of each treatment immediately after harvest. Averages followed by the standard error of the mean (SEM) (*n* = 6).

Harvest Treatment	Harvest Date	Brix	pH	Titratable Acidity (g/L)
First harvest	3-Feb	21.16 ± 0.08 a [a]	3.24 ± 0.18 a	8.18 ± 0.15 c
Second harvest	27-Feb	23.14 ± 0.09 b	3.48 ± 0.01 b	5.05 ± 0.04 b
Third harvest	29-Mar	25.16 ± 0.08 c	3.75 ± 0.01 c	4.61 ± 0.04 a

[a] Means followed by different letters within the same column indicate significant differences for Fisher's least significant difference (LSD) test and *p* < 0.05.

3.2. Basic Chemical Composition of the Finished Wines

Table 2 reports the basic chemical composition of the 18 wines produced as a result of the combination of harvesting Merlot fruit at three maturity levels and applying MW, in addition to Control (i.e., untreated) treatments (three harvest treatments × two winemaking treatments × triplicate fermentations). Table 2 shows a two-way ANOVA presenting *p*-values for the separate effects of harvest time, the two winemaking treatments, and their interaction. The harvest treatment clearly impacted the basic chemistry of the wines, whereas the winemaking treatments did not. Differences in alcohol content between harvest treatments reflected differences previously observed as a function of grape maturity in Brix levels (Table 1). The pH of the wines also mirrored fruit pH showing the expecting trend of wine pH being higher than fruit pH, although, for the wines of the third harvest, wine pH was lower than fruit pH. This discrepancy for the wines of the third harvest was probably due to an error at the moment of sampling. Winemakers often report that the actual fruit pH can only be effectively assessed once the crushed fruit, that is must, has soaked in contact with the solids for, at least, 24 h post-crushing. It is possible that in the third harvest fruit, not all the acids were released upon crushing, therefore leading to lower than expected titratable acidity and higher than expected pH. Only alcohol content was affected as a result of the winemaking treatments, with MW-treated wines of the second harvest showing higher levels relative to the corresponding Control wines from this harvest.

The acid composition of the wines, broadly defined by titratable acidity and encompassing tartaric, citric, malic, and lactic acid, as well as volatile acids (acetic acid) was generally not affected by the winemaking treatments. However, the harvest treatment did have an impact on these parameters. Of note was lower tartaric acid levels in the wines of the second harvest and lower citric acid levels in the wines of the third harvest. Malolactic fermentation was not completed in both Control and MW wines of the second harvest, as evidenced by the residual malic acid content in the vicinity of 0.90 to 0.93 g/L malic acid. Malolactic fermentation was only partially completed in the wines of the other two harvests.

Regarding the carbohydrate composition of the wines, both the content of Glu+Fru, and the residual sugar content, were generally higher in the wines of the third harvest. However, differences in residual sugars with the wines of the first and second harvest are unlikely to be of chemical and/or sensory relevance.

Mirroring trends observed for Brix (in the fruit) and alcohol content (in the finished wines), the glycerol content of the wines, although unaffected by the winemaking techniques, increased progressively from the first to the third harvest, with the wines of the first harvest showing the lowest glycerol content and the wines of the third harvest showing the highest, with wines of the second harvest placing in between. This result is consistent with previous reports indicating increasing concentrations of glycerol in wines made from grapes of increasing sugar levels [9,23].

Table 2. Two-way analysis of variance (ANOVA) of the basic chemical composition of Merlot wines of the different harvest and microwave treatments at bottling. Averages followed by the standard error of the mean (SEM) ($n = 3$).

Harvest Treatment	Microwave Treatment	Alcohol % (v/v)	pH	Titratable Acidity (g/L)	Tartaric Acid (g/L)	Citric Acid (g/L)	Malic Acid (g/L)	Lactic Acid (g/L)	Acetic Acid (g/L)	Glucose + Fructose (g/L)	Residual Sugars (g/L)	Glycerol (g/L)
First harvest	Control	12.41 ± 0.06 a	3.43 ± 0.09 a	5.57 ± 0.03 b	2.32 ± 0.03 d	0.31 ± 0.05 b	0.50 ± 0.06 a	1.53 ± 0.03 bc	0.92 ± 0.01 b	0.60 ± 0.28 a	0.87 ± 0.21 a	9.50 ± 0.01 a
	MW	12.42 ± 0.08 a	3.46 ± 0.07 a	5.68 ± 0.12 b	2.22 ± 0.05 cd	0.27 ± 0.08 b	0.57 ± 0.03 a	1.51 ± 0.07 bc	0.93 ± 0.03 b	0.98 ± 0.41 a	1.26 ± 0.66 a	9.87 ± 0.26 ab
Second harvest	Control	13.50 ± 0.06 b	3.60 ± 0.07 cd	5.53 ± 0.17 b	1.73 ± 0.10 a	0.37 ± 0.05 b	0.93 ± 0.06 b	1.49 ± 0.06 bc	0.93 ± 0.03 b	0.27 ± 0.14 a	0.73 ± 0.35 a	10.43 ± 0.07 bc
	MW	13.76 ± 0.09 c	3.62 ± 0.02 d	5.50 ± 0.06 b	1.60 ± 0.05 a	0.41 ± 0.08 b	0.90 ± 0.06 b	1.54 ± 0.11 c	0.92 ± 0.01 b	0.47 ± 0.06 a	0.53 ± 0.27 a	10.41 ± 0.15 bc
Third harvest	Control	15.11 ± 0.06 d	3.58 ± 0.09 bc	5.08 ± 0.12 a	2.11 ± 0.06 bc	0.05 ± 0.02 a	0.50 ± 0.11 a	1.33 ± 0.03 ab	0.81 ± 0.01 a	1.93 ± 0.35 b	2.81 ± 0.11 b	10.56 ± 0.17 c
	MW	15.27 ± 0.09 d	3.54 ± 0.01 b	5.13 ± 0.07 a	2.03 ± 0.04 b	0.06 ± 0.03 a	0.43 ± 0.03 a	1.18 ± 0.06 b	0.89 ± 0.02 b	2.01 ± 0.32 b	3.16 ± 0.33 b	10.83 ± 0.32 c
ANOVA factors and interactions												
Harvest treatment (H)		**<0.0001** [b]	**<0.0001**	**0.001**	**<0.0001**	**<0.0001**	**<0.0001**	**0.002**	**0.001**	**<0.0001**	**<0.0001**	**0.001**
Microwave treatment (MW)		**0.025**	0.830	0.612	**0.034**	0.827	0.831	0.476	0.116	0.387	0.535	0.229
H × M interaction		0.194	0.056	0.805	0.842	0.745	0.555	0.317	0.092	0.874	0.658	0.575

[a] Means followed by different letters within the same column indicate significant differences for Fisher's LSD test and $p < 0.05$. [b] Significant p values ($p < 0.05$) are shown in bold fonts.

3.3. Phenolic and Chromatic Composition of the Wines

Figure 1 shows the extraction and evolution of selected phenolic classes and color during the length of winemaking, through malolactic fermentation, and after a short aging period, spanning a total of 150 days after crush. One of the main goals of the present study was to follow the effect of MW-assisted extraction throughout aging. This is because other similar extraction techniques based on heating crushed grapes, including thermomaceration and flash-détente, have been reported to produce a rapid increase in phenolic compounds, which tends to dissipate shortly after alcoholic fermentation [24] or during aging [25]. Additionally, Figure 2 presents the achieved improvements in phenolic compounds and color, when applicable, on a relative basis both at pressing and after 150 days post-crushing. The chromatic and phenolic composition of Control wines was considered as a baseline to establish these relative percentages of extraction upon MW treatment presented in Figure 2.

The most noticeable effect of MW was the fast increase in the wines' anthocyanin content upon applying this treatment to the musts of the different harvests. For example, in the case of unripe fruit, the freshly crushed must of Control wines had 180 mg/L of anthocyanins, whereas MW caused a quick extraction, amounting to 560 mg/L, that is, a 211% improvement in anthocyanin content. An equally enhancing effect on anthocyanin extraction upon crushing was observed in MW musts of the second harvest. However, in the case of the ripest fruit at 25.1 Brix, the application of MW only increased anthocyanins by 89% relative to Control wines upon crushing. These results suggest that MW favored anthocyanin extraction early during winemaking in the case of unripe fruit, but this effect diminished when MW was applied to fully ripe fruit. Ripening of *Vitis vinifera* L. grapes encompasses complex physiological and biochemical processes, whereby skin cell walls and vacuoles undergo progressive degradation and cells become increasingly disorganized as ripening progresses, thereby, favoring the extraction of vacuolar components, such as anthocyanins and tannins [26]. Ripening also causes de-pectination and de-esterification of the grape berry cell walls, which further phenolic extractability into wine [27]. The fruit of the third harvest was fully ripe, both from a commercial and technical viewpoint. Thus, it is possible that phenolic extractability was naturally high in this fully ripe fruit, with further de-pectination of cell walls occurring during fermentation. Consequently, applying MW to this fruit caused little improvement. In fact, Figure 1 suggests that, indeed, anthocyanins were easily extractable in the fully ripe fruit, as these wines were generally higher in anthocyanins than the wines from the first and second harvest throughout winemaking. Figure 2 confirms that both at pressing, and to a lesser extent after 150 days post-crushing, anthocyanin extraction was improved more in the wines of the first harvest than in the wines of the second or third harvest.

Tannins are polymers of monomeric flavan-3-ols units and are located in both skins and seeds. Although seed- and skin-derived tannins differ in their concentration and composition, tannins are broadly defined by their ability to bind and precipitate proteins [28]. This biochemical fact explains both why seed and skin tannins can elicit astringency (which occurs as a result of the precipitation of salivary proline-rich proteins in human saliva by wine tannins) and why tannins can be measured by precipitating them with a standardized protein such as bovine serum albumin (BSA) [21]. The present study reports this fraction of wine tannins, that is, the vast majority of them that precipitate with proteins and are able to elicit astringency. In the measurement performed just upon crushing, the application of MW caused little effect on the initial tannin extraction in the case of the wines of the first and second harvest. Effects on the wines of both harvests were seen after day 6 (first harvest) and after day 17 (second harvest). The lack of early tannin extraction in these wines may be due to the source from which the bulk of the tannin material was extracted from during maceration. Skin tannins are vacuolar compounds that are relatively easier to extract in the earlier stages of maceration [29]. However, on a berry mass basis, skin tannins are less abundant than seed tannins for most *Vitis vinifera* L. varieties [30]. Seed tannins are confined in thin-walled cells below the seed cuticle [5]. As a result, the extraction of seed tannins into wine is more difficult and progresses at a slower rate relative to that of skin tannins. Because red wines are produced by allowing the contact with solids for a relatively prolonged period of time, seed-derived tannins account for the majority of tannins extracted into wine, particularly when extended maceration is favored [31,32]. In the present study, wines were allowed a 14-day maceration period, but it is not possible to assert the origin of the tannins extracted into these wines. A previous study on the application of another thermal technique, flash-détente, to red wines, showed that this technique favored the extraction of skin-derived tannins over that of seed-derived tannins [13]. However, the precise underlying mechanisms behind MW-assisted extraction are unknown, and, thus, the relative extractions rates achieved by MW may differ from those achieved by flash-détente or thermonaceration.

In the case of the fruit of the third harvest, their respective wines were generally higher in tannins, more so in the case of MW-treated wines. As this fruit was slightly overripe, it is possible that skin and seed tannin extractability was also higher, thereby explaining both the early increase and the long-term positive effect on wine tannin content. Another possibility is the enhancing role of ethanol on tannin extraction in the case of overripe fruit, which generated higher alcohol levels. Overall, tannin extraction upon MW treatment applied to unripe fruit increased by 30% at pressing, but only improvements of 20 and 10% were seen for the wines made from fruit of the second and third harvest, respectively (Figure 2).

Total phenolics are measured upon reaction with iron-chloride, with Fe^{3+} reacting with all phenolics containing vicinal dihydroxyls, but not with monohydroxylated phenols and anthocyanins [33]. Therefore, the measurement of total phenolics does include tannins, but also flavan-3-ols and flavonols. At day 0 and upon crushing, MW treated musts of the first harvested saw a 250% increase in total phenolics relative to their respective Control wines. Improvements of lesser magnitude upon application of MW were seen in the case of the fruit of the second and third harvest. At pressing, improvements or around 34% in total phenolics were recorded in MW-treated wines of the first harvest, whereas these improvements were of 20 and 12% for the MW-treated wines of the second and third harvest, respectively (Figure 2). The maximum levels of total phenolics were achieved in the last sampling point at day 150 post-crushing, with equivalent levels in wines made from unripe and fully ripe fruit in the case of MW-treated wines. At day 150 post-crush, MW-treated wines showed significantly higher levels of total phenolics than Control wines irrespective of fruit maturity at harvest.

Total polymeric pigments encompass the summation of small polymeric pigments (SPP), which do not precipitate proteins, and large polymeric pigments (LPP), which they do. SPP are composed of low-molecular weight pigments, including pyranoanthocyanins and flavan-3-ol-anthocyanin acetaldehyde-mediated adducts [34]. LPP are covalent adducts between tannins and anthocyanins of relatively high molecular weight, which can elicit astringency, but also confer long-term color stability [34]. For simplicity, Figure 1 reports the evolution of total polymeric pigments, that is, SPP + LPP, throughout maceration and early aging of the full set of wines.

Polymeric pigment formation was minimally or not affected at all by the MW treatment in the case of the wines from the first and second harvest. However, application of the MW treatment in the wines of the third harvest produced an initially higher formation of polymeric pigments. During aging, polymeric pigment formation was favored in MW treated wines of only the first and third harvest. In effect, polymeric pigment formation was unaffected by MW in the wines of the second harvest. In the case of wines made from unripe (first harvest) or fully ripe (third harvest) fruit, higher polymeric pigment content may be explained by higher tannin extraction in these wines. In the wines of the second harvest, conversely, tannin extraction upon MW treatment was only marginally favored. It has been shown that LPP formation is favored in conditions of relatively higher availability of tannins [31]. This was certainly the case of the wines of the first harvest, and, secondarily, that of the wines of the third harvest as well.

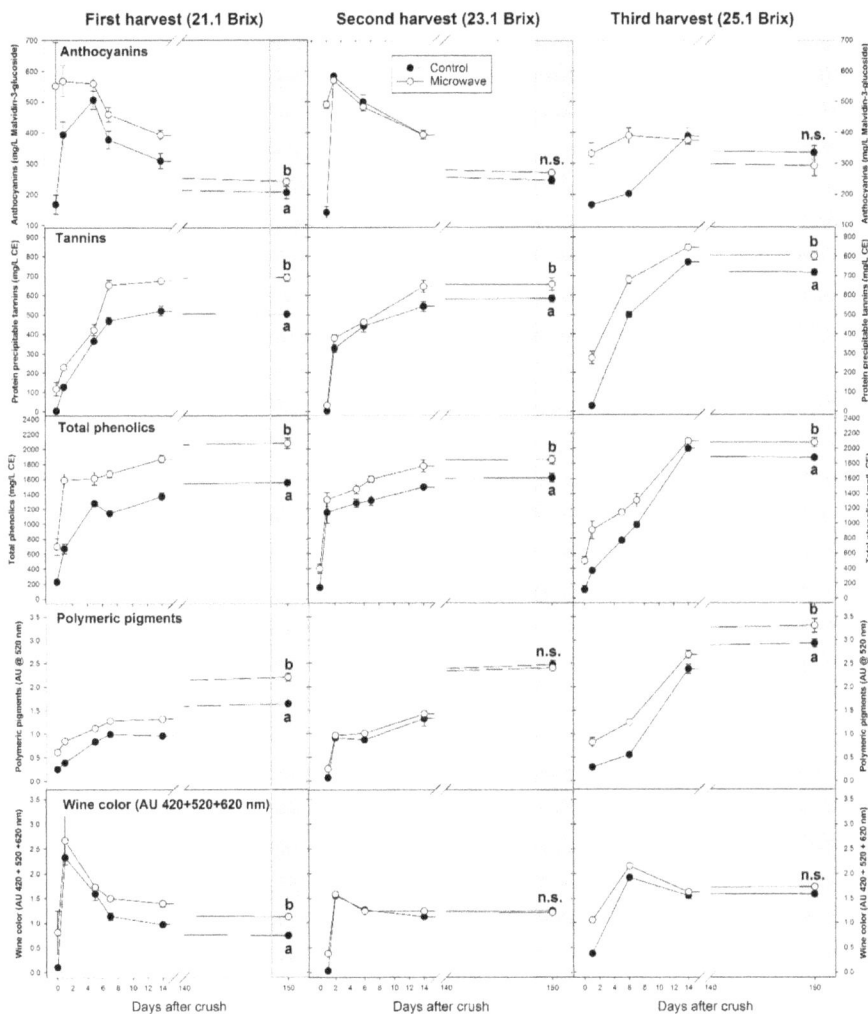

Figure 1. Evolution of anthocyanins, tannins, total phenolics, polymeric pigments and wine color in Merlot wines made from fruit harvested with three maturity levels and processed without ("Control") and with microwave-assisted extraction ("Microwave"). Different letters at day 150 post-crush between treatments made from fruit the same maturity level indicate significant differences for Student's *t*-tests and $p < 0.05$. CE: catechin equivalents; AU: absorbance units. n.s.: not significant.

It was also observed that anthocyanin extraction increased upon MW treatment in the wines made from unripe fruit. At pressing, in accordance with the observed improved anthocyanin extraction, MW treated wines of the first harvest showed a 48% improvement in SPP. Improvements of much lesser magnitude in SPP were observed for the wines of the second and third harvest. Improvements in LPP upon application of MW were of a relatively lower magnitude to those observed for SPP, and they became even negative in the case of wines of the second harvest treated with MW (Figure 2). Overall, these results suggest that the MW treatment preferentially favored the formation of polymeric pigments in unripe fruit and this improvement hinged upon enhanced extraction of anthocyanins. Small polymeric pigments generally correlate with initial higher content of free anthocyanins [2].

This explains the favored formation of SPP over LPP in the MW treated wines of the first harvest. This finding further provides support to the early statement that MW treatment favored the extraction of anthocyanins (but not that of seed tannins) in unripe fruit and, thus, eventually favored the formation of SPP over LPP.

Notably, when all the three harvests were considered, the polymeric content of the wines increased progressively along with the ripeness of the fruit from which they were made of. In other words, the polymeric pigment content was higher in the wines made from fully ripe fruit at 25.1 Brix. A study in Syrah in which fruit was harvested at 21, 23, and 25 Brix also reported a higher concentration of SO_2-resistant pigments (equivalent to polymeric pigments), in wines made from fruit harvested at 23 and 25 Brix [35]. Similar trends were reported by Bindon and colleagues in Cabernet Sauvignon, whereby riper fruit produced wines with a higher concentration of SO_2-resistant pigments relative to unripe fruit, which the authors attributed to higher tannin content in the wines made from riper fruit [9].

Wine color, herein determined as the summation of absorbances at 420 nm (yellow hues), 520 nm (red hues), and 620 nm (blue hues) was followed during winemaking (Figure 1). Hue (420 nm/520 nm) was also determined, and it is presented as a percentage increase or decrease relative to Control wines (Figure 2). These color parameters are routinely used by wineries and are referred to as the "Glories color indexes" [20]. Initial color improvements as a result of the application of the MW treatment in the fruit from the three harvests were of higher magnitude in the case of the fruit of the second harvest (23.1 Brix) (300% initial improvement) than in unripe fruit (21.1 Brix) (275%), and fully ripe fruit (25.1 Brix) (175%). However, in the wines of the second harvest treated with MW, this initial difference disappeared shortly after the onset of alcoholic fermentation and remained equivalent with the color measured in Control wines throughout winemaking. This was also the case of wines of the third harvest wherein initial improvements in wine color of MW-treated wines decreased shortly after pressing and completely converged with Control wines by day 150 post-crushing. In the case of unripe fruit, however, initial differences in favor of MW-treated wines were maintained during winemaking, in coincidence with previously reported improvements in anthocyanins, tannins, total phenolics, and polymeric pigments in these wines. Wine color improvements in MW-treated wines of the first harvest amounted to 45% at pressing and 52% at day 150 post-crushing. Furthermore, hue decreased upon application of MW in these wines, indicating these wines had more bluish and less yellow hue in their color composition, a sign of slower color evolution relative to Control wines (Figure 2). As mentioned above, improvements of wine color in MW-treated wines of the second and third harvest were only marginal at pressing and disappeared completely or even turned negative by day 150 post-crushing. Accordingly, the hue values were generally higher in MW-treated wines of the second and third harvest relative to Control wines.

In summary, phenolic content and color increases in wines made from unripe fruit treated with MW ranged from 23 to 48% at pressing and from 18 to 50% after 150 days post-crushing. These improvements were particularly noticeable in the case of wine color (Figure 2). Improvements of lesser magnitude on phenolic content and color were recorded in the case of MW-assisted extraction applied to fruit of the second and third harvest.

A closer analysis of Figures 1 and 2 suggest a harvest treatment × winemaking treatment interaction wherein for most phenolic and color parameters the effect of the MW treatment yielded better results when applied to unripe fruit. In other words, the outcomes of the MW treatment seemed contingent upon grape maturity.

To explore this hypothesis, the phenolic and chromatic composition of the wines were analyzed by a two-way ANOVA keeping the separate effects of the harvest treatment, the winemaking treatments, and their interaction, both at pressing and after 150 days post-crushing (Tables S1 and S2, respectively). Apart from the anthocyanin content of the wines at pressing, the remaining phenolic and chromatic parameters were affected by the harvest treatments at pressing and at day 150 post-crushing. This result is not surprising, as several reports exist on the effect of harvest date on red wine phenolic composition and color [3,9,35]. The application of MW also affected the phenolic and chromatic composition of the wines, but this was more evident at day 150 post-crushing (Table S2). This result indicates that the effect of the MW treatment, unlike that of thermomaceration or flash-détente, is relatively long-lasting.

Interestingly, analysis of the harvest treatment × winemaking treatment interaction at pressing only indicated a significant interaction for anthocyanins (Table S1); the remaining phenolic and chromatic parameters being affected similarly by the MW treatments irrespective of grape maturity. However, at day 150 post-crushing, LPP, total polymeric pigments, tannins, and total phenolics all showed significant harvest treatment × winemaking treatment interactions. This suggests that the long-term effect of MW did depend upon harvest time. That is that MW-assisted extraction applied to unripe fruit produced positive and long-lasting effects on the phenolic and chromatic composition of these wines. Since the most critical changes in phenolics and color are relatively confined to maceration and early aging, including the period between the end of alcoholic fermentation and that of malolactic fermentation [4], the results herein presented provide a valid snapshot of the short-term evolution of phenolics and color of these wines.

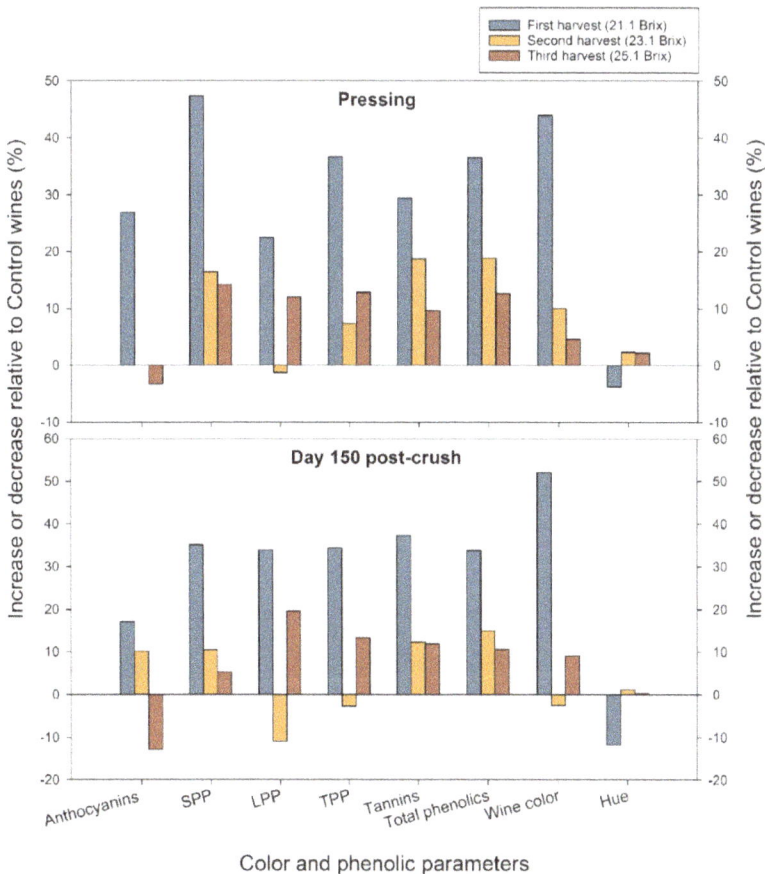

Figure 2. Percentages of increase/decrease of color and phenolic parameters in Merlot wines produced with MW-assisted extraction relative to Control wines at pressing (day 14) and day 150 post-crushing. SPP: small polymeric pigments; LPP: large polymeric pigments; TPP: total polymeric pigments; wine color (AU 420 nm + 520 nm + 620 nm); hue (AU 420 nm/520 nm).

3.4. Detailed Anthocyanin and Anthocyanin-Derived Pigment Composition of the Wines

Anthocyanins and anthocyanin-derived pigments were measured after bottling by HPLC-DAD and peak identity was confirmed by mass spectroscopy (Table S3). A total of 26 anthocyanin and anthocyanin-derived pigments were identified in these Merlot wines. To facilitate the interpretation of the results, pigments were grouped as glycosides (i.e., monoglucosylated anthocyanins, including delphinidin-, cyanidin-, petunidin-, peonidin-, and malvidin-3-glucosides); acetylated (i.e., acylated anthocyanin monoglucosides, with the acyl moiety being acetic acid); coumaroylated (i.e., acylated anthocyanin monoglucosides, with the acyl moiety being *p*-coumaric acid); pyranoanthocyanins; tannin-anthocyanin dimers of direct condensation or mediated by acetaldehyde; and total pigments (i.e., the overall summation of all pigmented molecules quantified) (Figure 3). Figure 3 also presents the results of a one-way ANOVA considering all the harvest and winemaking treatments together. The concentration of each of these pigment families increased concurrently with fruit ripeness, that is, higher concentrations of each pigment families were generally found in the wines made from the riper fruit at 25.1 Brix. In the wines made from less ripe fruit, that is, 21.1 and 23.1 Brix, the MW treatment enhanced anthocyanin and anthocyanin-derived pigment contents, but this trend was not seen in the case of the wines of the third harvest. To confirm these observations, a series of Student's *t*-tests were carried out comparing the Control and MW treatments of each harvest individually. In the wines of the first harvest, MW increased the total pigment content of the treated wines by 19% relative to Control wines ($p = 0.031$), as well as that of pyranoanthocyanins ($p = 0.006$) and of anthocyanin-tannin dimers ($p = 0.008$). In the wines of the second harvest, this same increase in MW-treated wines was of 4% relative to Control wines, but was not statistically significant ($p = 0.319$). No differences were found between Control and MW-treated wines for the remaining pigment families. Likewise, no differences between Control and MW wines were found for any of the pigment families in the wines of the third harvest. These results are consistent with previous trends observed for tannins, total phenolics, and wine color whereby enhancing effects of the MW treatment were observed in the wines made from unripe fruit.

Interestingly, the MW treatment caused a positive effect in the content of pyranoanthocyanins and anthocyanin-tannin dimers in the wines of the first harvest. Pyranoanthocyanins are anthocyanin-derived pigments formed by reaction of monomeric anthocyanins with acetaldehyde, acetoacetic acid, pyruvic acid and other carbonyl compounds [36]. Anthocyanin-tannin dimers are formed via covalent reactions between anthocyanins and tannins and can be mediated by acetaldehyde [37]. While both pyrantoanthocyanins and anthocyanin-tannin dimers provide stable color, they only accounted for 10 and 3% on a concentration basis, respectively, of the total pigment content of the wines of the first harvest. Furthermore, both the results of the one-way ANOVA and independent Student's *t*-tests confirmed that, irrespective of fruit maturity, MW did not alter the extraction of monoglucosylated and acylated (acetyladed + coumaroylated) anthocyanins, which were quantitatively the major pigment families in these wines. This implies that MW generally favored anthocyanin extraction in unripe fruit but did not favor the extraction and retention of any major pigments over others. Finally, it is worth mentioning that this work did not determine the anthocyanin content of the fruit. This analysis would have allowed gauging the effectiveness of the MW treatment on anthocyanin extraction and retention into wine as a function of fruit maturity. Further work considering the initial fruit phenolic composition and the efficiency of MW in extracting specific phenolic compounds is currently underway.

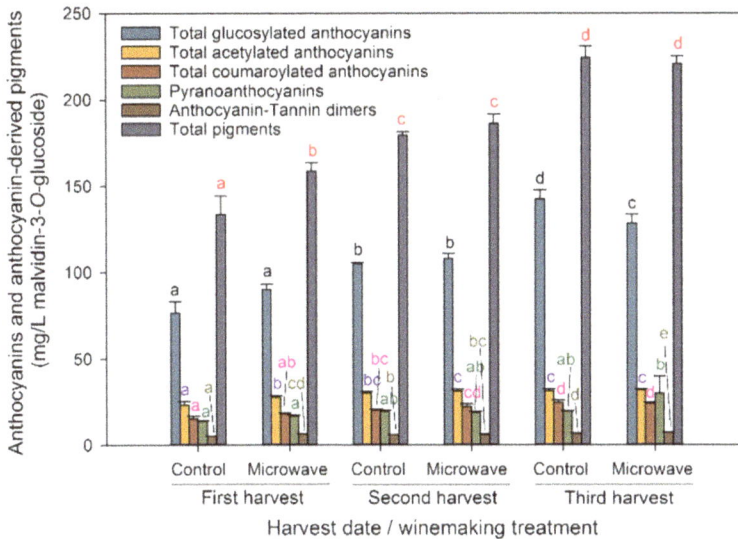

Figure 3. Detailed anthocyanin composition of Merlot wines made from fruit harvested with three maturity levels and processed without ("Control") and with microwave-assisted extraction ("Microwave") at bottling. Different letters for a given pigment group within treatments indicate significant differences for Fisher's least significant difference (LSD) and $p < 0.05$. To facilitate comprehension, letters bearing the same color are associated with one specific pigment group.

4. Conclusions

In this work, unripe, ripe, and fully ripe Merlot grapes were harvested and made into wine prior application of MW-assisted extraction. Untreated Control wines were also produced, with winemaking procedures being kept consistent within harvest dates and throughout winemaking.

Results suggest that the MW treatment produced positive improvements on anthocyanins, tannins, total phenolics, polymeric pigments, and color particularly when this treatment was applied to unripe Merlot fruit harvested with 21.1 Brix. There was a lesser effect of MW on these same parameters on the wines made from ripe (23.1 Brix) and fully ripe (25.1 Brix) fruit, and when early improvements were observed, these generally disappeared during winemaking. In the case of fully ripe fruit, the application of MW caused negative effects in some phenolic compounds. This is consistent with the reported effect of similar heat-based techniques, such as thermomaceration or flash-détente, on red wines wherein early improvements in phenolics and color tend to subside after alcoholic fermentation.

The positive results observed upon application of MW to unripe fruit suggests that this technique furthers phenolic extraction in the case of fruit that may be deficient in phenolic maturity. As such, MW can become a valid tool to handle fruit that is either unripe or may be compromised by fungal or other diseases, as well as healthy fruit that may necessitate early harvest due to logistic reasons. In the latter case, application of MW would allow the production of highly colored wines of moderate or relative lower alcohol content. These results need to be validated for other varieties and viticultural conditions. The feasibility to scale up the MW technology to industrial scale for handling crushed grapes requires subsequent evaluation as well.

Supplementary Materials: Supplementary materials can be found at http://www.mdpi.com/2311-5637/5/1/15/s1.

Author Contributions: L.F.C. conceived the research, designed and performed the experiments, performed the statistical analysis, evaluated the results, and wrote the paper; S.E.S. performed the experiments, analyzed the

Fermentation **2019**, *5*, 15

samples, and evaluated the results; E.A.B. performed the experiments; M.L.F. analyzed the samples, performed the statistical analysis and evaluated the results. All the authors read and approved the final version of the manuscript.

funding: This research was funded by the Instituto Nacional de Tecnología Agropecuaria (INTA) Grant Number MZASJ-1251101. INTA is thanked for financial and logistical support of the project.

Acknowledgments: INTA personnel are thanked for help during harvest and winemaking. Vanesa Garcia is acknowledged for skillful analytical support.

Conflicts of Interest: The authors declare no conflict of interest.

References

1. Bindon, K.; Holt, H.; Williamson, P.O.; Varela, C.; Herderich, M.; Francis, I.L. Relationships between harvest time and wine composition in Vitis vinifera L. cv. Cabernet Sauvignon 2. Wine sensory properties and consumer preference. *Food Chem.* **2014**, *154*, 90–101. [CrossRef] [PubMed]
2. Casassa, L.F. Phenolic management in red wines: Investigation of the timing and severity of Regulated Deficit Irrigation (RDI), grape maturity and selected maceration conditions by HPLC-MS and sensory techniques. Ph.D. Thesis, Washington State University, Pullman, WA, USA, 2013.
3. Casassa, L.F.; Beaver, C.W.; Mireles, M.; Larsen, R.C.; Hopfer, H.; Heymann, H.; Harbertson, J.F. Influence of fruit maturity, maceration length, and ethanol amount on chemical and sensory properties of Merlot wines. *Am. J. Enol. Vitic.* **2013**, *64*, 437–449. [CrossRef]
4. Casassa, L.F.; Harbertson, J.F. Extraction, evolution, and sensory impact of phenolic compounds during red wine maceration. *Annu. Rev. Food Sci. Technol.* **2014**, *5*, 83–109. [CrossRef] [PubMed]
5. Adams, D.O. Phenolics and Ripening in Grape Berries. *Am. J. Enol. Vitic.* **2006**, *57*, 249–256.
6. Chen, J.-Y.; Wen, P.-F.; Kong, W.-F.; Pan, Q.-H.; Wan, S.-B.; Huang, W.-D. Changes and subcellular localizations of the enzymes involved in phenylpropanoid metabolism during grape berry development. *J. Plant Physiol.* **2006**, *163*, 115–127. [CrossRef] [PubMed]
7. Casassa, L.F. Flavonoid Phenolics in Red Winemaking. In *Phenolic Compounds—Natural Sources, Importance and Applications*; Soto-Hernandez, M., Palma-Tenango, M., Garcia-Mateos, M.d.R., Eds.; InTech: Rijeka, Croatia, 2017. [CrossRef]
8. Gil, M.; Kontoudakis, N.; González, E.; Esteruelas, M.; Fort, F.; Canals, J.M.; Zamora, F. Influence of Grape Maturity and Maceration Length on Color, Polyphenolic Composition, and Polysaccharide Content of Cabernet Sauvignon and Tempranillo Wines. *J. Agric. Food Chem.* **2012**, *60*, 7988–8001. [CrossRef]
9. Bindon, K.; Varela, C.; Kennedy, J.; Holt, H.; Herderich, M. Relationships between harvest time and wine composition in Vitis vinifera L. cv. Cabernet Sauvignon 1. Grape and wine chemistry. *Food Chem.* **2013**, *138*, 1696–1705. [CrossRef]
10. Setford, P.C.; Jeffery, D.W.; Grbin, P.R.; Muhlack, R.A. Factors affecting extraction and evolution of phenolic compounds during red wine maceration and the role of process modelling. *Trends Food Sci. Technol.* **2017**, *69*, 106–117. [CrossRef]
11. Koyama, K.; Goto-Yamamoto, N.; Hashizume, K. Influence of Maceration Temperature in Red Wine Vinification on Extraction of Phenolics from Berry Skins and Seeds of Grape (*Vitis vinifera*). *Biosci. Biotechnol. Biochem.* **2007**, *71*, 958–965. [CrossRef]
12. Doco, T.; Williams, P.; Cheynier, V. Effect of Flash Release and Pectinolytic Enzyme Treatments on Wine Polysaccharide Composition. *J. Agric. Food Chem.* **2007**, *55*, 6643–6649. [CrossRef]
13. Morel-Salmi, C.; Souquet, J.-M.; Bes, M.; Cheynier, V. Effect of Flash Release Treatment on Phenolic Extraction and Wine Composition. *J. Agric. Food Chem.* **2006**, *54*, 4270–4276. [CrossRef] [PubMed]
14. Aguilar, T.; Loyola, C.; de Bruijn, J.; Bustamante, L.; Vergara, C.; von Baer, D.; Mardones, C.; Serra, I. Effect of thermomaceration and enzymatic maceration on phenolic compounds of grape must enriched by grape pomace, vine leaves and canes. *Eur. Food Res. Technol.* **2016**, *242*, 1149–1158. [CrossRef]
15. Carew, A.L.; Gill, W.; Close, D.C.; Dambergs, R.G. Microwave Maceration with Early Pressing Improves Phenolics and Fermentation Kinetics in Pinot noir. *Am. J. Enol. Vitic.* **2014**, *65*, 401–406. [CrossRef]
16. Chandrasekaran, S.; Ramanathan, S.; Basak, T. Microwave food processing—A review. *Food Res. Int.* **2013**, *52*, 243–261. [CrossRef]

17. Casassa, L.F.; Huff, R.; Miller, E. Effect of Stem Additions and Microwave Extraction of Musts and Stems on Syrah, Merlot, and Cabernet Sauvignon Wines. In Proceedings of the 68th American Society for Enology and Viticulture National Conference, Seattle, WA, USA, 26–29 June 2017.

18. Carew, A.L.; Sparrow, A.; Curtin, C.D.; Close, D.C.; Dambergs, R.G. Microwave Maceration of Pinot Noir Grape Must: Sanitation and Extraction Effects and Wine Phenolics Outcomes. *Food Bioprocess Technol.* **2013**. [CrossRef]

19. Iland, P.; Bruer, N.; Edwards, G.; Caloghiris, S.; Wilkes, E. *Chemical Analysis of Grapes and Wine Techniques and Concepts*, 2nd ed.; Patrick Iland Wine Promotions Pty Ltd.: Adelaide, Australia, 2012; p. 118.

20. Glories, Y. La couleur des vins rouges. 2ème partie. Mesure, origine et interprétation. *Connaissance de la Vigne et du Vin* **1984**, *18*, 253–271.

21. Harbertson, J.F.; Picciotto, E.A.; Adams, D.O. Measurement of polymeric pigments in grape berry extracts and wines using a protein precipitation assay combined with bisulfite bleaching. *Am. J. Enol. Vitic.* **2003**, *54*, 301–306.

22. Blanco-Vega, D.; López-Bellido, F.J.; Alía-Robledo, J.M.; Hermosín-Gutiérrez, I. HPLC–DAD–ESI-MS/MS Characterization of Pyranoanthocyanins Pigments Formed in Model Wine. *J. Agric. Food Chem.* **2011**, *59*, 9523–9531. [CrossRef]

23. Rankine, B.C.; Bridson, D.A. Glycerol in Australian Wines and Factors Influencing Its Formation. *Am. J. Enol. Vitic.* **1971**, *22*, 6–12.

24. El Darra, N.; Turk, M.F.; Ducasse, M.-A.; Grimi, N.; Maroun, R.G.; Louka, N.; Vorobiev, E. Changes in polyphenol profiles and color composition of freshly fermented model wine due to pulsed electric field, enzymes and thermovinification pretreatments. *Food Chem.* **2016**, *194*, 944–950. [CrossRef]

25. de Andrade Neves, N.; de Araújo Pantoja, L.; dos Santos, A. Thermovinification of grapes from the Cabernet Sauvignon and Pinot Noir varieties using immobilized yeasts. *Eur. Food Res. Technol.* **2014**, *238*, 79–84. [CrossRef]

26. Robinson, S.P.; Davies, C. Molecular biology of grape berry ripening. *Aust. J. Grape Wine Res.* **2000**, *6*, 175–188. [CrossRef]

27. Garrido-Bañuelos, G.; Buica, A.; Schückel, J.; Zietsman, A.J.J.; Willats, W.G.T.; Moore, J.P.; Du Toit, W.J. Investigating the relationship between cell wall polysaccharide composition and the extractability of grape phenolic compounds into Shiraz wines. Part II: Extractability during fermentation into wines made from grapes of different ripeness levels. *Food Chem.* **2019**, *278*, 26–35. [CrossRef] [PubMed]

28. Hagerman, A.E.; Butler, L.G. The specificity of proanthocyanidin-protein interactions. *J. Biol. Chem.* **1981**, *256*, 4494–4497. [PubMed]

29. Peyrot des Gachons, C.; Kennedy, J.A. Direct Method for Determining Seed and Skin Proanthocyanidin Extraction into Red Wine. *J. Agric. Food Chem.* **2003**, *51*, 5877–5881. [CrossRef] [PubMed]

30. Harbertson, J.F.; Kennedy, J.A.; Adams, D.O. Tannin in skins and seeds of Cabernet Sauvignon, Syrah, and Pinot Noir berries during ripening. *Am. J. Enol. Vitic.* **2002**, *53*, 54–59.

31. Casassa, L.F.; Beaver, C.W.; Mireles, M.S.; Harbertson, J.F. Effect of extended maceration and ethanol concentration on the extraction and evolution of phenolics, colour components and sensory attributes of Merlot wines. *Aust. J. Grape Wine Res.* **2013**, *19*, 25–39. [CrossRef]

32. Harbertson, J.F.; Mireles, M.S.; Harwood, E.D.; Weller, K.M.; Ross, C.F. Chemical and sensory effects of saignée, water addition, and extended maceration on high Brix must. *Am. J. Enol. Vitic.* **2009**, *60*, 450–460.

33. Harbertson, J.F.; Spayd, S. Measuring Phenolics in the Winery. *Am. J. Enol. Vitic.* **2006**, *57*, 280–288.

34. Adams, D.O.; Harbertson, J.F.; Picciotto, E.A. Fractionation of red wine polymeric pigments by protein precipitation and bisulfite bleaching. In *Red Wine Color*; American Chemical Society: Washington, DC, USA, 2004; Vol. 886, pp. 275–288.

35. Garrido-Bañuelos, G.; Buica, A.; Schückel, J.; Zietsman, A.J.J.; Willats, W.G.T.; Moore, J.P.; Du Toit, W.J. Investigating the relationship between grape cell wall polysaccharide composition and the extractability of phenolic compounds into Shiraz wines. Part I: Vintage and ripeness effects. *Food Chem.* **2019**, *278*, 36–46. [CrossRef]

36. de Freitas, V.; Mateus, N. Formation of pyranoanthocyanins in red wines: A new and diverse class of anthocyanin derivatives. *Anal. Bioanal. Chem.* **2011**, *401*, 1463–1473. [CrossRef] [PubMed]
37. Fulcrand, H.; Dueñas, M.; Salas, E.; Cheynier, V. Phenolic Reactions during Winemaking and Aging. *Am. J. Enol. Vitic.* **2006**, *57*, 289–297.

fermentation

MDPI

Review

Enzymes for Wine Fermentation: Current and Perspective Applications

Harald Claus [1,*] and Kiro Mojsov [2]

[1] Institute of Molecular Physiology, Microbiology and Wine Research, Johannes Gutenberg-University, 55099 Mainz, Germany

[2] Faculty of Technology, University "Goce Delčev" Štip, Krste Misirkov No. 10-A, P.O. Box 201, 2000 Štip, Macedonia; kiro.mojsov@ugd.edu.mk

* Correspondence: hclaus@uni-mainz.de

Received: 16 May 2018; Accepted: 5 July 2018; Published: 9 July 2018

Abstract: Enzymes are used in modern wine technology for various biotransformation reactions from prefermentation through fermentation, post-fermentation and wine aging. Industrial enzymes offer quantitative benefits (increased juice yields), qualitative benefits (improved color extraction and flavor enhancement) and processing advantages (shorter maceration, settling and filtration time). This study gives an overview about key enzymes used in winemaking and the effects of commercial enzyme preparations on process engineering and the quality of the final product. In addition, we highlight on the presence and perspectives of beneficial enzymes in wine-related yeasts and lactic acid bacteria.

Keywords: wine clarification; extraction; pectinase; glycosidase; protease; phenoloxidase; color; aroma; non-*Saccharomyces* yeasts

1. Introduction

Over the last decades, commercial enzyme preparations have gained increasing popularity in the wine industry [1–5]. They offer many advantages such as accelerated settling and clarification processes, increased juice yield, and improved color extraction (Table 1).

Table 1. Enzymes used for winemaking and their function.

Application/Process	Enzymatic Activity	Aim
Enhancement of filtration/clarification of must	Pectinolytic enzymes	Degradation of viscosity (pectin)
Mash fermentation/heating (red wine)	Pectinase with side activities (cellulases, hemicellulases)	Hydrolysis of plant cell wall polysaccharides. Improvement of skin maceration and color extraction of grapes, quality, stability, filtration of wines
Late phase of fermentation (white wine)	Glycosidases	Improvement of aroma by splitting sugar residues from odorless precursors
Young wine	Glucanases	Lysis of yeast cell walls, release of mannoproteins
Contaminated juice	Glucanases	Lysis of microbial exopolysacharides to improve clarification
Wine	Urease	Hydrolysis of yeast derived urea, preventing formation of ethyl carbamate
Must, wine	Lysozyme from hen egg	Control of bacterial growth
Must, wine	Proteases	Wine stabilization by prevention of protein haze; Reduction of bentonite demand

Technical enzyme preparations are usually obtained from fungi, which are cultured under optimal conditions on substrates to facilitate their preparation and purification. In contrast to grape-derived enzymes, which are often inactive under wine conditions (low pH, presence of ethanol, phenolic compounds, sulphite etc.), fungal enzymes are resistant. The production of oenological enzymes for use in the European Union is regulated by the International Organization of Vine and Wine (OIV), which has ruled that *Aspergillus niger* and *Trichoderma* sp. may be used as source organisms (i.e., have GRAS, "generally regarded as safe" status) [1,2]. Selected strains from *A. niger* are fermented under aerobic conditions in optimized growth media for production of pectinases, hemicellulases and glycosidases, *Trichoderma* species are used for production of glucanases and *Lactobacillus fermentum* for urease.

Current commercial enzyme preparations are usually cocktails of different activities, such as glucosidases, glucanases, pectinases and proteases [5]. The search for enzymes with improved and more specific characteristics will continue. In this respect, the study and exploration of the high endogenous enzyme potential of wine and grape-associated microorganisms (Table 2) will assist the wine industry to meet prospective technical and consumer challenges. In contrast to filamentous fungi, yeasts with beneficial enzymatic endowment, could be directly used as starter cultures, without application of enzyme preparations.

Table 2. Microbial enzymes with relevance for winemaking.

Enzyme	Remarks
Fungi (*Botrytis cinerea*)	
Glycosidases	Influence aromatic potential of infected grapes by release of volatile aroma compounds
Laccases	Broad specificity to phenolic compounds, cause oxidation and browning
Pectinases	Depolymerizing enzymes, cause degradation of plant cell walls and grape rotting
Cellulases	Multi-component complexes: endo-, exoglucanases and cellobiases; synergistic working, degrade plant cell walls
Lipases	Degrade lipids (e.g., in cell membranes)
Esterases	Involved in ester formation and degradation
Proteases	Aspartic proteases occur early in fungal infection, determine rate and extent of rotting caused by pectinases
Yeasts	
Glucosidases	Some yeasts produce β-glucosidases which are not repressed by glucose and are resistant to ethanol and low pH; positive influence on wine flavor
Glucanases	Occur extracellular, cell wall associated and intracellular, accelerate autolysis process and release of mannoproteins
Proteases	Acidic endoproteases accelerate autolysis process and degradation of grape proteins
Pectinases	Degrade pectin in grape cell walls
Lactic acid bacteria	
Malolactic enzymes	Convert malic acid to lactic acid
Esterases	Involved in ester formation and degradation
Glycosidases	Deliberate flavor compounds
Lipases	Degrade lipids
Lichenases, Glucanases, Cellulases, Xylanases	Degradation of polysaccharides
Proteases	Hydrolysis of proteins
Tannases	Hydrolysis of tannins (polymeric phenolic compounds)
Laccases	Oxidation of phenolic compounds

In the following paragraphs we give an overview on widely used enzyme preparations for wine fermentation and a special focus on wine-associated microorganisms as alternative enzyme sources.

2. Pectinases

The grape cell wall consists of cellulose microfibrils linked together by a matrix of xyloglucan, mannan, xylan (hemicellulose) and pectin, all of which is stabilized by a protein network. The high viscosity of pectin, which is dissolved after berry crushing impedes juice extraction, clarification and filtration. In addition, pectin prevents diffusion of phenolic and aroma compounds into the must during wine fermentation.

2.1. Commercial Pectinases

The complete degradation of pectin needs cooperation of several enzymes to break the complex molecule into small fragments [1,2]. They include different enzymatic activities:

- Polygalacturonase (homogalacturonan-hydrolase) (PG): hydrolytic depolymerization of the polygalacturonic acid chain. One can differentiate enzymes that cleave either single galacturonic acid units from the chain end (exo-activity, exoPG, EC 3.2.1.67), or in the middle of the chain (endo-activity, endoPG, EC 3.2.1.15).
- Pectinlyase/pectate lyase (EC 4.2.2.2 and 4.2.2.9): nonhydrolytic cleavage of the polygalacturonic acid chain.
- Pectinesterase (EC 3.1.1.11): hydrolytic cleavage of methanol from the D-galacturonic acid chain, causing drastic viscosity reduction in the liquid portion of the mash and better must flow.
- Acetylesterase (EC 3.1.1.6): cleaves acetyl residues from D-galacturonic acid with release of acetic acid. By this way the interfering acetyl residues at the connecting points of the side chains of the "hairy regions" are removed which facilitates further enzymatic degradation.

Most commercial preparations are derived from fungal sources [1–5] and are more or less well-defined enzyme mixtures (Table 3). The application of bulk enzyme preparations is advantageous as it fulfills several functions. Examples are liquefaction enzymes, which contain cellulases and hemicellulases in addition to pectinases.

Commercially available pectinase preparations contain the active enzymes (2–5%) and additives (sugars, inorganic salts, preservatives) which stabilize and standardize the specificities of the products [1]. Factors that generally inhibit proteins will reduce effectiveness of the enzymes. These include juice clarification with bentonite, which adsorbs and deposits the proteins. Alcohol levels above 17% (v/v) and SO_2 levels above 500 mg/L inhibit pectinases [4]. Tannin-rich wines show reduced enzyme activity as phenolic polymers react with the proteins and render them useless.

Table 3. Examples of commercial pectinase preparations used for winemaking modified from [3,4].

Supplier	Enzyme Preparation	Purpose of Application
AEB, South Africa	Pectocel L	Improvement of clarification, filtration and product yield
	Endozym Pectoflot	Improvement of clarification, filtration and product yield
	Endozym Contact Pelliculaire	Enhancement of extraction and color stabilization
	Endozym Rouge	Enhancement of extraction and color stabilization
	Endozyme Active	Improvement of clarification, filtration and product yield
Begerow, Germany	Siha Panzym Extract G	Enhanced extraction and release of color and aroma
	Siha Panzym Clair Rapide G	Improvement of clarification, filtration and product yield
	Siha Panzym Fino G (β-Glucanase)	Improvement of clarification, filtration and sensory
	Siha Panzym Arome G (β-Glucosidase)	Enhanced aroma development
Darleon, South Africa	Influence	Improvement of clarification, filtration and product yield
	Enzym' Color Plus	Enhancement of extraction and color stabilization
DSM, Switzerland	Rapidase Filtration	Improvement of clarification, filtration and product yield
	Rapidase Vino Super	Improvement of clarification, filtration and product yield
Enartis, Italy	Utrazym 1000S	Clarification of white juices—facilitation of fining and filtration
	Progress Quick	Must flotation
	Utrazym couleur	Enhanced extraction during short macerations
Erbslöh, Germany	Trenolin bukett DF (β-Glycosidase)	Enhanced aroma development—Improvement of clarification
	Trenolin Super DF	Improvement of clarification, filtration and product yield
	Trenolin Flot DF	Must flotation
	Trenolin 4000 DF	Enhancement of sugar yield
	Trenolin Filtro DF (β-Glucanase)	Improvement of clarification and filtration; Hydrolysis of *Botrytis cinerea* exopolysaccharide slime
	Trenolin Bukett DF	Enhance of color and aroma release from red grapes
Laffort, France	Lafazym press	Enhanced color and tannin extraction—Facilitation of fining and filtration
	Lafazym CL	Improvement of clarification, filtration and product yield
	Lafase 60	Improvement of clarification, filtration and product yield
	Lafase HE	Enhancement of extraction and color stabilization
Lallemand, France	Lallemand EX	Enhancement of extraction and color stabilization
	Lallemand OE	Enhancement of extraction and color stabilization
Novo Nordisk, Denmark	Novoclair FCE	Improvement of clarification, filtration and product yield
	Vinozym EC	Enhancement of extraction and color stabilization
	Glucanex (β-Glucanase)	Improvement of clarification and filtration; Hydrolysis of *Botrytis cinerea* exopolysaccharide slime
	Ultrazym	Improvement of clarification and filtration
	Pectinex Superpress	Improvement of clarification and filtration
Valley Research, USA	Crystalzyme	Rapid clarification—Color improvement—Increased complexity—Process efficiency

2.1.1. Effect on Juice Extraction, Clarification and Filtration

The pulp of the grape varieties is rich in pectin compounds. The incomplete hydrolysis of these molecules by the endogenous enzymes can therefore cause problems in processing. If pectinases are applied to the pulp prior to pressing, they can improve juice and color yield. For the clarification of musts after pressing, pectinase-based enzyme preparations are recommended. Its pectin methyl esterase and endogalacturonic activities cause hydrolysis of the pectin chains and facilitate the drainage of juice from the pomace with an increased yield of a free-flowing juice with lower viscosity [6–8]. In addition, it causes cloud particles to aggregate into larger units that deposit as sediment. The acceleration of the clarification process also produces more compact lees. When applied to the pulp before pressing, it increases juice yield and color yield [6–8].

2.1.2. Effect on Color Extraction

Anthocyanidins are the red grape pigments, which mainly occur in the grape skin [9]. The chemical structure, commonly referred to as "flavylium cation", is characterized by two benzene rings linked by an unsaturated cationic heterocycle. Normally, the dye molecule is linked to a glucose monomer, which improves water solubility and stability. Pelargondin, cyanidin, delphinidin, peonidin, petundin and malvidin are the main variants identified in grapes and wine.

Flavonols are light yellow pigments found in the skins of both red and white grapes [9]. These are mainly the glycosylated forms of kaempferol, quercetin and myricetin. In red wine concentrations are in the range of 100 mg/L, in white wines 1–3 mg/L.

Under natural conditions, solubilization of phenolic compounds from grapes is facilitated by increased ethanol concentrations in the course of alcoholic fermentation. However, the extraction is uncomplete as the grape skin forms a physical barrier against the diffusion of anthocyanins, tannins and flavors from the cells. Therefore, various oenological techniques have been developed that result in wines that have good visual characteristics and are as stable as possible [10]. Especially wines made by pectinase treatment showed higher concentrations of anthocyanins and total phenols, as well as greater color intensity and optical clarity compared to untreated control wines [6–8].

2.1.3. Immobilization

Immobilization is a commonly used strategy to conserve the desirable properties of enzymes for biotechnological applications. In addition to improved stability, immobilization offers a number of advantages, such as reusability, ease of product separation, and better control of catalysis. Various methods have been described for the immobilization of pectinases, such as inclusion in alginate [11], physical adsorption to anionic resins [12], and covalent bonding to supports such as porous glass [13] and nylon [14]. A pectinase from *Aspergillus niger* immobilized on chitosan-coated carriers retained 100% of its original activity after several cycles of reuse [15].

2.2. Yeast Pectinases

S. cerevisiae strains, despite their genetic ability to secrete an endo-polygalacturonase, usually show no or only weak pectinase activity. In contrast, many so-called "wild" yeasts have been identified to be pectinase producers [16–20].

Grape fermentations at low temperatures (15–20 °C) are believed to protect the volatile compounds, thereby improving the aromatic profile of the wines. Therefore, cold-active enzymes are required for both extraction and clarification [21]. Psychrophilic yeasts are natural sources for such biocatalysts [22]. The pectinolytic enzymes of *Cystofilobasidium capitatum* and *Rhodotorula mucilaginosa* are effective under oenological pH (3.5) and temperature conditions (6.0 °C and 12 °C). Also, pectinases from several *A. pullulans* strains remain active at wine-relevant concentrations of glucose, ethanol or SO_2, bearing the potential as processing aids for low-temperature wine fermentations [23].

3. Lipases

Lipids in wine originate directly from the grape berry [24] and by autolysis of wine yeast [25]. It has been reported that lipid composition undergoes considerable changes during wine fermentation [26].

Authentic lipases (E.C. 3.1.1.3) are mainly active at the oil–water interface of emulsified substrates with long fatty acid chains. Triglycerides are cleaved rendering glycerol and fatty acids. In contrast, carboxylic ester hydrolases (see below) hydrolyze soluble esters with relatively short fatty acid chains [27]. However, the transition between both activities appears somewhat fluid.

Lipolytic activities have been detected only in few wine-relevant *Lactobacillus* strains [28], but in different genera of yeasts isolated from natural environments [29,30]. In theory, lipases could be used for winemaking for decomposition of lipoid cell membranes, thereby improving color extraction from red grape berries (Figure 1).

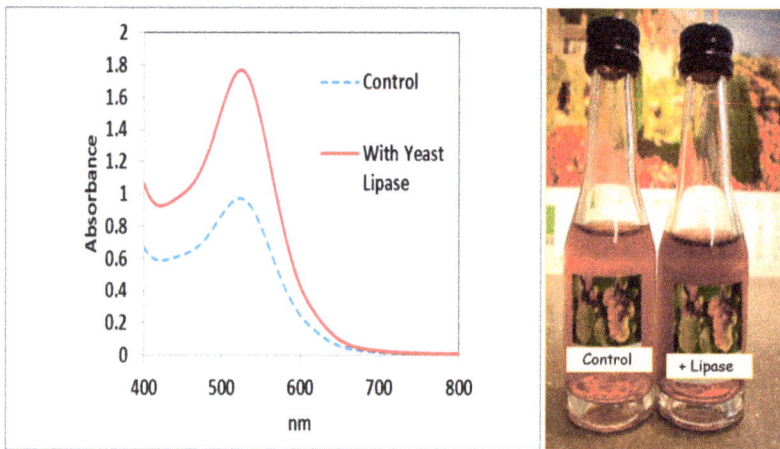

Figure 1. Color extraction from red grapes after a model fermentation without and in presence of a yeast lipase (Claus, unpublished data).

4. Glucanases

4.1. Commercial Glucanases

Polysaccharides in must and wine originate from the grape berries (cellulose, hemicellulose, pectins) and from cell walls of yeasts during growth and autolysis (beta-glucans, chitin). Several strains of lactic acid bacteria (especially *Pediococcus* spp.) and the grape fungus *Botrytis cinerea* produce viscous capsular or extracellular polysaccharides impairing wine filtration [31]. The colloidal polysaccharides cannot be removed from wine by flocculants, adsorbents or filtration. Thus, commercial products with glucanase activities e.g., those from *Trichoderma* sp., and *Taleromyces versatilis* are useful to reduce viscosity of musts and wine caused by microbial contaminations [5] (Table 3).

Two types of glucanases are relevant for wine fermentation: (i) exo-β-1,3-glucanases split β-glucan chains by sequentially cleaving glucose residues from the non-reducing end and releasing glucose as the sole hydrolysis product and, (ii) endo-β-1,3-glucanases catalyse the intramolecular hydrolysis of β-glucans with release of oligosaccharides.

Some yeast cell constituents, in particular the wall, can exert a significant impact on the technological and sensory properties of wine. The cell wall consists of β-glucans (~60%), mannoproteins (~40%) and chitin (~2%). In particular, the mannoprotein fraction has attracted increasing interest in wine fermentation to stabilize tartaric acid and protein, improve mouthfeel and reduce astringency [32–34]. The use of mannoproteins of the cell wall and chitin as binding elements for the removal of undesirable compounds such as ochratoxin A and toxic heavy metals has been proposed. The mannoprotein fractions present in the yeast cell walls are highly variable between species and even strains, providing opportunities for the development of alternative fining products that could replace conventional proteinaceous animal preparations with allergenic potential.

Commercial enzyme preparations (Table 3) facilitate the release of these components. As a prerequisite they have to be effective under acid wine conditions, moderate temperatures and in presence of ethanol [35].

4.2. Microbial Glucanases

A major microbial source of polysaccharide-degrading exoenzymes are non-*Saccharomyces* yeasts belonging to the genera *Kloeckera*, *Candida*, *Debaryomyces*, *Rhodotorula*, *Pichia*, *Zygosaccharomyces*, *Hanseniapora*, *Kluyveromyces* and *Wickerhamomyces* (Table 4). Glucanolytic enzyme activities were also detected in wine-relevant lactic acid bacteria [28] (Table 2).

Table 4. Glucanases of non-*Saccharomyces* yeasts with possible use for winemaking modified from [4,18].

Species	Specificity	Substrate	MW(kDa)
Candida albicans	Endo-β-1,3-	L, OL, P	49
	Exo-β-1,3-	L	107
Candida hellenica	nd	G	nd
Candida lambica	nd	G	nd
Candida pulcherrima	nd	G, Li	nd
Candida stellata	nd	G, Li	nd
Candida utilis	Endo-β-1-3-	L, PNPG	20
	Exo-β-1,3-1,6-	L, P, PNPG	20
	Endo-β-1,3-	L, OL	21
Kloeckera apiculata	nd	G, Li	nd
Kluyveromyces phaseolosporus	Endo-β-1,3-(I)	L, OL	180
	Exo-β-1,3-(II)	L, OL	45
	Exo-β-1,3-1,6-(III)	L, P	18.5
	Exo-β-1,3-1,6-(IV)	L, Ol, P	8.7
Pichia polymorpha	Endo-β-1,3-(I)	L, OL	47
	Exo-β-1,3-1,6-(II)	L, OL, P, PNPG	40
	Exo-β-1,3-(III)	L, PNPG	30
Schizosaccharomyces pombe	Endo-β-1,3-(I)	L, OL	160
	Endo-β-1-3-(II)	L, OL	75
Schizosaccharomyces versatilis	Endo-β-1,3-	L, OL	97
	Exo-β-1,3-1,6-	L, P, PNPG	43
Wickerhamomyces anomalus AS1	Exo-β-1,3-	L, PNPG	47.5

L: laminarin, OL: oxidized laminarin, P: pustulan, G: β-glucan (Barley), Li: lichenan, PNPG: p-nitro-phenyl-β-D-glucoside; nd: not determined.

5. Glycosidases

The organoleptic properties of wine are determined by a variety of different compounds that are already present in the grape (aroma) or arise during fermentation or storage (bouquet). Acids such as tartaric acid or citric acid affect the taste, but the characteristic odor and taste is mainly due to volatile organic substances such as esters, alcohols, thiols or terpenes [36–38].

Due to their low odor threshold, particularly terpenes determine wine flavor. Like other aroma active compounds (C13 norisoprenoids, benzene derivatives, aliphatic alcohols, phenols), they are secondary metabolites mainly derived from the grape skin. Approximately 90% of these compounds do not exist in a free form, but are conjugated to mono- or disaccharides, thereby forming water-soluble and odourless complexes. The most frequently occurring aroma precursors in grape varieties such as *Muscat* and *Riesling* are the glycosidic bound terpenes linalool, nerol and geraniol.

The sugar residues consist of rutinoside (rhamnose-glucose), arabinoside (arabinose-glucose) or apioside (apiose-glucose) [38].

Enzymatic hydrolysis of sugar-conjugated precursors release very aromatic, volatile terpenes (aglycones). Usually, the terminal sugars are first cleaved off by a rhamnosidase, an arabinosidase or an apiosidase. In a second step, the terpenes are released by a β-D-glucopyranosidase. This means that the latter activity alone can release only terpene compounds bound to a single glucose residue. In addition to a stepwise reaction, some glucosidases are able to hydrolyze the glycosidic bond to the aglycone, regardless of the number of sugar residues [39,40]. Nowadays commercial enzyme preparations are available that can hydrolyze the disaccharide directly from the terpene in a single step [5]. In many bulk enzyme preparations, glycosidase activities occur as side activities along with pectinase and glucanase activities (Table 3).

An important microbial source of wine-related enzymatic activities are lactic acid bacteria [28] (Table 2). Perez-Martin et al. [41] studied >1000 isolates for glycosidases. The β-glucosidase activities were only found in cells, but not in the supernatants of the cultures. Four *O. oeni* isolates retained their enzymatic activity under the conditions of winemaking. In a similar study, cell-bound glucosidase and arabinosidase activities from *O. oeni* strains released high levels of monoterpenes from natural substrates under optimal conditions [42]. The enzymes showed broad substrate specificities (release of both primary and tertiary terpene alcohols) and remained active in grape juice.

Glycosidase activities have been also detected in various non-*Saccharomyces* yeasts (*Candida*, *Hanseniaspora*, *Pichia*, *Metschnikowia*, *Rhodotorula*, *Trichosporon*, *Wickerhamomyces*) [43–52] (Table 5). Several experiments on the technical application of yeast glycosidases to improve organoleptic quality of wines gave positive results [40,44,46,49,50].

Polyphenols in red wine, such as resveratrol, have gained increasing public and scientific interest due to their supposed beneficial effects on human health [53]. A majority of the polyphenols in nature are conjugated to sugars or organic acids, making them more hydrophilic and less bioavailable to humans. The amount of glycosylated forms of resveratrol, known as piceid or polydatin, has been found to be up to ten times higher in red wines. Since these modified forms are less bioactive, experiments with β-glucosidases from various fungal sources have been undertaken to increase the trans-resveratrol content in wines by hydrolysis of glycosylated precursors. The multifunctional glucanase WaExg2 of *W. anomalus* AS1 released the aglycones from the model compounds arbutin, salicin, esculin and polydatin [52]. WaExg2 was active under typical wine conditions such as low pH (3.5–4.0), high sugar concentrations (up to 20% *w/v*), high ethanol concentrations (10–15% *v/v*), presence of sulphites and various cations. Therefore, this yeast strain could be useful in wine production for several purposes: to increase the levels of sensory and beneficial compounds by cleaving glycosylated precursors or reducing the viscosity by hydrolysis of glycan slurries. In this context Madrigal et al. [29] underlined that glucose- and ethanol-tolerant enzymes from *Wickerhamomyces* are of great interest to the wine industry.

Table 5. Aroma enhancing enzymes of non-*Saccharomyces* yeasts with possible use for wine fermentation modified from [54].

Species	Enzymatic Activities*				
	β-D-Glucosidase	α-L-Arabino-Furanosidase	α-L-Rhamnosidase	β-D-Xylosidase	Carbon-Sulfur Lyase
Aureobasidium pullulans	+	+	+		
Brettanomyces anomalus	+				
Candida guillermondii	+		+	+	
Candida molischiana	+				
Candida stellata	+		+	+	
Candida utilis				+	
Candida zemplinia					+
Debaryomyces castelli	+				
Debaryomyces hansenii	+				
Debaryomyces polymorphus	+				

Table 5. *Cont.*

Species	Enzymatic Activities*				
	β-D-Glucosidase	α-L-Arabino-Furanosidase	α-L-Rhamnosidase	β-D-Xylosidase	Carbon-Sulfur Lyase
Debaryomyces pseudopolymorphus	+				
Debaryomyces vannjii	+				
Hanseniaspora guillermondii	+				
Hanseniaspora osmophila	+			+	
Hanseniaspora vineae	+	+	+	+	
Hanseniaspora uvarum	+	+	+	+	
Issatschenkia terricola	+				
Kluyveromyces thermotolerans	+				+
Metschnikowia pulcherrima	+			+	+
Pichia angusta			+		
Picha anomala	+	+	+	+	
Pichia capsulata	+	+			
Pichia guilliermondii			+		
Pichia kluyveri					+
Pichia membranaefaciens	+			+	
Saccharomycodes ludwigii	+				
Schizosaccharomyces pombe	+				
Sporidiobolus pararoseus	+				
Torulasporus delbrueckii	+				+
Torulasporus asahii	+				
Wickerhamomyces anomalus	+	+		+	
Zygosaccharomyces bailii	+				

* Activity detected (+).

6. Esterhydrolases and -Synthetases

Esters (e.g., ethyl acetate, isoamyl acetate, ethyl hexanoate, ethyl octanoate, ethyl decanoate) contribute to the most desirable fruity wine flavors [36,37]. They are synthesized by the grapes but are also produced by yeasts in course of alcoholic fermentation [55]. During malolactic fermentation, significant changes in the concentration of individual esters were observed [56]. Presence of alcohol acyltransferases (ester synthesis) and esterases (ester hydrolysis) in wine yeasts [37] and lactic acid bacteria [57] is well documented.

Depsides are esters of aromatic hydroxycarboxylic- or phenolic acids with each other or with other carboxylic acids of the grape, such as tartaric acid [5]. These compounds can be hydrolyzed by cinnamoyl esterases ("depsidases"), which often appear as side activities in enzyme preparations made from *A. niger* [5] The fission products can have a negative influence on wine quality. Enzymatic deliberated phenolcarboxylic acids, such as caffeic acid or coumaric acid, can be converted by the yeast metabolism to the volatile phenol derivatives 4-vinylguajacol and 4-vinylphenol, which are unpleasant side-tastes in the wine [5]. Therefore, commercial pectinase preparations should be free of depsidase side activities. In a recent study of 15 commercial enzyme preparations, approximately half of the samples yielded significant cinnamoyl esterase activities [58].

7. Proteinases

Proteins in must in wine are derived from the grapes, and from microbial cells (yeasts, lactic, acid bacteria) and their activities [59]. Another important source are protein-based wine additives (e.g., lysozyme, ovalbumin, gelatin, casein) which could pose allergenic-like reactions to consumers [60–62]. Most of these proteins have vanished after termination of wine fermentation and subsequent fining procedures. However so-called pathogen-related (PR) proteins (β-glucanases, chitinases, thaumatin-related proteins) can still be present. They are synthesized by the plants for defence against bacterial or fungal infections and in response to abiotic stress [63]. Due to their compact structures, they are resistant against acid wine conditions, heat, and proteolysis [59].

In combination with other wine ingredients, PR proteins can cause undesirable turbidity especially during cold storage of white wines with negative economic consequences [59]. Currently, protein removal is mainly achieved by bentonite addition [64], a process that can be associated with decreased wine quantity and quality. Bentonite acts essentially as a cation exchanger, and individual wine proteins adsorb to different degrees on the clay [64]. Proteins that are negatively charged at wine pH (about 3.5) and/or are highly glycosylated as laccases of *Botrytis cinerea* are less bound by bentonite. Thus, new fining agents are desired to remove proteins from wine.

7.1. Proteases from Fungal and Plant Sources

Enzymatic degradation of wine proteins seems an attractive alternative to bentonite treatment as it would minimize losses of volume and aroma. As a prerequisite, suitable proteases have to be active under specific wine conditions (acid pH, presence of ethanol, sulphites, phenolics) and preferably act at low temperatures. Another challenge is the resistance of PR proteins against proteolysis due to their special molecular features like disulfide-bonds and glycosylations. Nevertheless, other grape proteins might be more susceptible, and thus proteases may help to reduce effective bentonite dosages. Currently, proteases from plants (papain, bromelain) have been tested with some promising results [65,66]. A fungal protease from *Aspergillus* sp. (aspergilloglutamic peptidase) has already approved for Australian winemaking [67]. The enzymatic procedure involves flash-pasteurization of grape must and is thus limited to specialized wineries. In this context, a protease of *Botrytis cinerea* BcAp8 has been described to hydrolyze grape chitinase at moderate temperatures [68].

7.2. Microbial Proteases

Microbial proteases can be an alternative or supplement to bentonite treatment for removal of unwanted wine proteins. Most *Saccharomyces cerevisiae* strains show no extracellular protease activity on diagnostic agar media [16–19,69]. However, a 72 kDa extracellular pepsin-like aspartic protease was characterized from a PIR 1 strain [70,71]. The enzyme was active during grape juice fermentations, although it did not affect turbidity-inducing proteins, unless the wine was incubated at 38 °C for extended time.

Proteinase A (PrA, saccharomycin; EC 3.4.23.25) is the major vacuole protease of *S. cerevisiae* encoded by the *PEP4* gene. As result of yeast autolysis, PrA enters wine in course of alcoholic fermentation. Far more, it has been found that under stress conditions (e.g., nutrient limitations) PrA is not targeted to the vacuole, but is misdirected to the cell membrane and secreted in the medium [72]. This would be advantageous for winemaking in view of haze reduction. Interestingly, the same situation is undesirable for beer brewery as PrA degrades proteins (e.g., lipid transfer protein 1), necessary for foam formation. Apart from PrA, *S. cerevisiae* expresses different cell-bound proteases, some of which are not fully characterized [73].

Non-*Saccharomyces* yeasts are important sources of extracellular enzymes including proteases (Table 6). Strains of *Metschnikowia pulcherrima* and *Wickerhamomyces anomalus* secreted aspartic proteases and degraded a model protein (bovine serum albumin) during growth in grape juice [74]. In a recent study, heterologous expressed aspartic protease MpAPr1 from *M. pulcherrima* [75] was added to a Sauvignon Blanc must. It was shown that the enzyme was active during fermentation and degraded wine proteins to some extent [76]. An alternative strategy would be to perform wine fermentations with appropriate protease-positive starter cultures. In addition to cost reductions, there are no administrative restrictions for yeast applications in must and wine, which must be taken into account with enzyme preparations.

Occurrence of proteolytic activities in lactic acid bacteria is also well-documented [28] (Table 2). Growth of *Oenococcus oeni* depends on the presence of amino acids in the culture medium because of deficiency in corresponding synthetic pathways. This bacterium secretes several proteases which may help to gain access to rare nitrogen sources during malolactic fermentation [77].

Table 6. Extracellular proteases of non-*Saccharomyces* yeasts with possible use for wine fermentation.

Species	Mode of Identification	Characterization	Reference
Candida apicola	Skim milk agar (pH 3.5), Gene Sequencing	Aspartic protease CaPR1 (39.2 kDa)	[78]
Candida stellata	Casein agar	nd	[18]
Hanseniaspora guelliermondii			
Hansenispora valbyenis	Casein agar and broth (pH 6.0)	nd	[79]
Hanseniapora occidentalis			
Hanseniaspora uvarum	Skim milk agar (pH 3.5)	nd	[16]
	Casein agar	nd	[18]
Kloeckera apiculata	Enzymatic; Inhibitor studies	Acid endopeptidase	[80]
Metschnikowia pulcherrima	Skim milk agar (pH 3.5)	nd	[16]
	Azocasein hydrolysis during fermentation of grape must	nd	[81]
	Skim milk agar (pH 3.5), Sequencing of protease gene, Purification	Aspartic protease pAPR1 (40.8 kDa)	[75,78]
	Casein agar	nd	[18]
	Skim milk agar (pH 4.5), Enzymatic; Inhibitor studies, LC-MS/MS	Aspartic protease	[74]
	Skim milk agar (pH 3.5)	nd	[16]
Wickerhamomyces anomalus	Skim milk agar (pH 4.5), Enzymatic; Inhibitor studies, LC-MS/MS	Aspartic protease WaAPR1 (47 kDa)	[74]

nd: not determined.

8. Phenoloxidases

Spontaneous and enzymatic oxidations exert dramatic effects on the final phenolic composition from the grape berry to bottled wine [82,83]. Once the integrity of the berries is destroyed, oxidative enzymes (phenoloxidases) and their phenolic substrates are exposed to the air, resulting in enzymatic browning (Figure 2).

Figure 2. Wine browning by *Botrytis*-laccase (right).

There are two classes of copper enzymes responsible for these reactions [82,83]: Tyrosinase (E.C. 1.14.18.1) hydroxylates monophenols to ortho-diphenols and oxidizes the latter to ortho-quinone intermediates which easily react further to polymeric, mostly colored products. Laccase (E.C. 1.10.3.2) has no monohydroxylase activity, and oxidizes a broad spectrum of different phenols and other compounds by a radical mechanism. Tyrosinase originates from grape berries [84], whereas laccases in must and wine are derived from epiphytic fungi, particularly *Botrytis cinerea* [82]. Phenolic compounds as caffeic acid, gallic acid, vanillic acid, ferulic acid, or especially resveratrol are known for their beneficial effects on human health. In addition to be radical scavengers, they are activators of the human's intrinsic cellular antioxidant system and have antimicrobial properties. Tyrosinase and laccase oxidize phenolic wine compounds and thus alter their antioxidant and antimicrobial properties [85–87]. In this context, it is an interesting observation that gallic acid oxidation by a laccase from *Trametes versicolor* was higher at 30 °C than at 45 °C. Although fungal laccases are usually more active at higher temperatures, the effect can be explained by reduced oxygen solubility under the experimental conditions [88].

Laccase is generally not very welcome in wine, but several studies have ruled out that controlled laccase treatments could promote wine stabilization and even improve sensory properties [89–93].

Volatile phenols particular produced by *Brettanomyces/Dekkera* sp. yeasts are associated with a serious "Brett" taste defect in wine. Lustrata et al. [94] used a laccase from *T. versicolor* to reduce concentrations of 4-ethylguaiacol and 4-ethylphenol in a synthetic model wine.

Biogenic amines (BA) are another class of undesirable compounds in wine [95–97]. They originate from the grape berries or are formed during fermentation by activities of decarboxylase-positive microorganisms [98–102]. Although more common in foods such as cheese, BA have received much attention in wine, as ethanol can enhance the negative effects on human health by inhibiting the enzymes responsible for the detoxification of these compounds [101].

Enzymatic degradation of BA is usually catalyzed by various classes of oxidases [103,104]. Depending on the type of prosthetic group, they can be classified into FAD-dependent (E.C. 1.4.3.4) and copper-containing amine oxidases (CAOs, E.C. 1.4.3.6). The latter have been detected in various yeasts such as *Kluyeromyces marxianus* or *Debaryomyces hansenii* [105,106]. These enzymes belong to the class of type 2 or "non-blue" copper proteins which convert primary amines to the corresponding aldehydes with an equimolar consumption of molecular oxygen and formation of hydrogen peroxide and ammonia.

Aromatic amines such as tyramine, phenylethylamine, tryptamine or serotonin are another class of compounds that can be oxidized by laccases [19]. Callejónet al. [107] detected enzymatic activities responsible for BA degradation in lactic acid bacterial strains isolated from wine. Responsible enzymes have been isolated and purified from *Lactobacillus plantarum* J16 and *Pediococcus acidilactici* CECT 5930 strains and have been identified as intracellular laccase-like multicopper oxidases. When the *L. plantarum* J16 laccase was overexpressed in *Escherichia coli*, it oxidized some BA, mainly tyramine [108].

9. Urease

Increased amounts of urea in wine can originate from yeast activities and then converted by a chemical reaction into the carcinogenic substance urethane (ethylcarbamate). During malolactic fermentation, lactic acid bacteria can produce other precursors of ethyl carbamate, such as arginine-derived citrulline and carbamyl phosphate. Especially at higher temperatures fermented wines may contain excessive amounts of urethane [109]. Therefore appropriate precautions should be taken to prevent the production of urethane. These include, for example, the selection of suitable starter cultures for malolactic fermentations and the reduction of arginine concentrations in the grape. Urease was introduced in 1997 by the EU as a new enzymatic wine treatment agent and can be used in exceptional cases. The enzyme splits urea into ammonia and carbon dioxide, preventing urethane formation. The commercial urease from *Lactobacillus fermentum* is effective on urea at doses of 50 mg/L in red wines and 25 mg/L in white wines [32].

10. Lysozyme

Yeasts, lactic acid and acetic acid bacteria have a significant influence on wine quality [110]. Microbial growth in musts and wines is conventionally controlled by the addition of sulfur dioxide. However, presence of sulphites in alcoholic beverages, particularly in wines, can cause pseudo-allergic responses with symptoms ranging from gastrointestinal problems to anaphylactic shock [32,33]. Other antimicrobials such as sorbic acid and dimethyl carbonate are primarily active against yeasts, but have limited activity against bacteria [32,34].

Lysozyme (EC 3.2.1.17) is a muramidase widely used to control microbial growth in foods such as cheese and wines [111,112]. Extensive enzymatic hydrolysis of the bacterial cell wall peptidoglycan, a polymer of N-acetyl-D-glucosamine units which are β-1,4-linked to N-acetylmuramic acid, results in cell lysis and death in hypoosmotic environments. Some lysozymes can kill bacteria by stimulating autolysin activity. In addition, bactericidal mechanisms involving membrane damage without enzymatic hydrolysis of peptidoglycan has been reported for c-type lysozymes, such as hen egg white lysozyme [113]. Gram-negative bacteria (i.e., acetic acid bacteria) are rather resistant against lysozyme, because the outer membrane acts as a barrier.

Lysozyme commercially produced from hen's egg white has been approved for winemaking by the International Organization of Vine and Wine in 2001 [114]. The amount added normally ranges between 250–500 mg/L. Four main applications and dosages are: (a) prevention of the onset of malolactic fermentation (early addition of 100–150 mg/L); (b) total inhibition of bacteria activity and malolactic fermentation (500 mg/L); (c) protection of wine during suboptimal alcoholic fermentation (250–300 mg/L); (d) stabilization of wine after malolactic fermentation (250–300 mg/L). Lysozyme can be eliminated by addition of fining agents, among which bentonite and metatartaric acid are the most efficient.

It has been reported that various Gram-positive strains of *Pediococcus* sp., *Lactobacillus* sp. and *Oenococcus oeni* [113] were not efficiently hydrolyzed by hen's egg white lysozyme. Reasonable explanations are structural modifications of the peptidoglycan, like N-deacetylation and O-acetylation of the glycan chains or amidation of free carboxyl groups of amino acids in the peptide chains [113]. As a possible alternative to hen's egg white lysozyme, exoenzymes (protease and muramidase) from *Streptomyces* species showed a broad bacteriolytic spectrum under winemaking conditions [98,113].

It should be mentioned that hen's egg lysozyme can display pH-dependent chitinase side activities. Under adverse conditions yeast cell walls (containing 2–4% chitin primarily in the bud scar regions) can be weakened by lysozyme with significant effects on vitality and stress response of *Saccharomyces cerevisiae* during wine fermentation [114].

11. Legislative Regulations

The use of enzymes in wine production in the European Union is regulated by the International Organisation of Vine and Wine (OIV). Specified resolutions define general aspects of enzymes in winemaking, the permitted enzyme activities, mode of application and enzyme activity measurements. The USA, Canada and China have national regulations in winemaking [5].

Genetic engineering. Today's enzyme production is based either on special selected wildtype strains or on genetically modified organisms (GMOs). The use of GMO production strains has considerable advantages: the product yields with GMOs is much higher than with wild strains and undesirable side activities become minimized. This makes it more efficient to produce and to guarantee the purity of the enzyme products. The labelling is regulated by the resolution OIV-OENO 485-2012.

GMOs for must fermentations. Although increasing numbers of *Saccharomyces* yeasts have been improved by genetically engineering [115,116], only two GMOs have been allowed for winemaking in three countries. The first recombinant strain to get official approval by appropriate food safety authorities (in the USA and Canada) was the malolactic wine yeast ML01. The GMO carries the *Schizosaccharomyces pombe* malate permease gene (*mae1*) and the *Oenococcus oeni* malolactic gene (*mleA*). The second strain ECMo01 expresses the urease gene constitutively to prevent formation of urethane [115]. Whether yeast strains obtained by protoplast fusion should be considered as GMO is in legal limbo.

12. Conclusions

Nowadays, the use of technical enzymes is a well-established strategy to improve wine quantity and quality. Currently they are mainly produced by *Aspergillus* species and applied as bulk preparations with several side activities. In view of consumer safety, more defined activities and alternative biological producers seem to be preferable. Yeasts, naturally occurring on grapes, have been found to be a rich source of oenological interesting enzymes. Their activities can be exploited in form of new enzyme products or directly as starter cultures for wine fermentation. This would satisfy the increasing trend to produce more individual wines with the aid of non-*Saccharomyces* yeasts [117].

Conflicts of Interest: The authors declare no conflict of interest.

References

1. Mojsov, K. Use of enzymes in wine making: A review. *Int. J. Market. Technol.* **2013**, *3*, 112–127.
2. Mojsov, K.; Andronikov, D.; Janevski, A.; Jordeva, S.; Zezova, S. Enzymes and wine—The enhanced quality and yield. *Adv. Technol.* **2015**, *4*, 94–100. [CrossRef]
3. Ugliano, M. Enzymes in winemaking. In *Wine Chemistry and Biochemistry*; Morena-Arribas, M.V., Polo, M.C., Eds.; Springer: New York, NY, USA, 2009; pp. 103–126.
4. Van Rensburg, P.; Pretorius, I.S. Enzymes in winemaking: Harnessing natural catalysts for efficient biotransformations: A review. *S. Afr. J. Enol. Vitic.* **2000**, *21*, 52–73.
5. Hüfner, E.; Haßelbeck, G. Application of microbial enzymes during winemaking. In *Biology of Microorganisms on Grapes, in Must and in Wine*, 2nd ed.; König, H., Unden, G., Fröhlich, J., Eds.; Springer International Publishing: Cham, Switzerland, 2017; pp. 635–658.
6. Mojsov, K.; Ziberoski, J.; Bozinovic, Z. The effect of pectolytic enzyme treatments on red grapes mash of Vranec on grape juice yields. *Perspect. Innov. Econ. Bus.* **2011**, *7*, 84–86. [CrossRef]
7. Mojsov, K.; Ziberoski, J.; Bozinovic, Z.; Petreska, M. A comparison of effects of three commercial pectolytic enzyme preparations in red winemaking. *Int. J. Pure Appl. Sci. Technol.* **2010**, *1*, 127–136.
8. Mojsov, K.; Ziberoski, I.; Bozinovic, Z.; Petreska, M. Effects of pectolytic enzyme treatments on white grape mashs of Smedervka on grape juice yields and volume of lees. In Proceedings of the 46th Croatian and 6th International Symposium on Agriculture, Opatija, Croatia, 14–18 February 2011; pp. 968–971.

9. Cosme, F.; Jordão, A.M. Grape phenolic composition and antioxidant capacity. In *Wine—Phenolic Composition, Classification and Health Benefits*; El Rayess, Y., Ed.; Nova Publishers: New York, NY, USA, 2014; pp. 1–40.

10. Sommer, S.; Cohen, S.D. Comparison of different extraction methods to predict anthocyanin concentration and color characteristics of red wines. *Fermentation* **2018**, *4*, 39. [CrossRef]

11. Reyes, N.; Rivas Ruiz, I.; Dominguez-Espinosa, R.; Solis, S. Influence of immobilization parameters on endopolygalacturonase productivity by hybrid *Aspergillus* sp. HL entrapped in calcium aliginate. *Biochem. Eng. J.* **2006**, *32*, 43–48. [CrossRef]

12. Sario, K.; Demir, N.; Acar, J.; Mutlu, M. The use of commercial pectinase in the fruit industry, part 2: Determination of the kinetic behaviour of immobilized commercial pectinase. *J. Food Eng.* **2001**, *47*, 271–274.

13. Romero, C.; Sanchez, S.; Manjon, S.; Iborra, J.L. Optimization of the pectinesterase/endo-D-polygalacturonase immobilization process. *Enzym. Microb. Technol.* **1989**, *11*, 837–843. [CrossRef]

14. Lozano, P.; Manjon, A.; Iborra, J.L.; Canovas, M.; Romojaro, F. Kinetic and operational study of a cross-flow reactor with immobilized pectolytic enzymes. *Enzym. Microb. Technol.* **1990**, *12*, 499–505. [CrossRef]

15. Ramirez, H.L.; Gómez Brizuela, L.; Úbeda Iranzo, J.; Areval-Villena, M.; Briones Pérez, A.I. Pectinase immobilization on a chitosan-coated chitin support. *J. Food Proc. Eng.* **2016**, *39*, 97–104. [CrossRef]

16. Charoenchai, C.; Fleet, G.H.; Henschke, P.A.; Todd, B.E.N. Screening of non-*Saccharomyces* wine yeasts for the presence of extracellular hydrolytic enzymes. *Aust. J. Grape Wine Res.* **1997**, *3*, 2–8. [CrossRef]

17. Fernández, M.; Úbeda, J.F.; Briones, A.I. Typing of non-*Saccharomyces* yeasts with enzymatic activities of interest in winemaking. *Int. J. Food Microbiol.* **2000**, *59*, 29–36. [CrossRef]

18. Strauss, M.L.A.; Jolly, N.P.; Lambrechts, M.G.; van Rensburg, P. Screening for the production of extracellular hydrolytic enzymes by non-*Saccharomyces* wine yeasts. *J. Appl. Microbiol.* **2001**, *91*, 182–190. [CrossRef] [PubMed]

19. Claus, H. Microbial Enzymes: Relevance for Winemaking. In *Biology of Microorganisms on Grapes, in Must and in Wine*, 2nd ed.; König, H., Unden, G., Fröhlich, J., Eds.; Springer International Publishing: Cham, Switzerland, 2017; pp. 315–338.

20. Mateo, J.J.; Maicas, S. Application of Non-*Saccharomyces* yeasts to winemaking process. *Fermentation* **2016**, *2*, 14. [CrossRef]

21. Merín, M.G.; Mendoza, L.M.; Farías, M.E.; de Ambrosini, V.I.M. Isolation and selection of yeast from wine grape ecosystem secreting cold-active pectinolytic activity. *Int. J. Food Microbiol.* **2011**, *147*, 144–148. [CrossRef] [PubMed]

22. Sahay, S.; Hanid, B.; Singh, P.; Ranjan, K.; Chauhan, D.; Rana, R.S.; Chaurse, V.K. Evaluation of pectinolytic activities for oenological uses from psychrotrophic yeasts. *Lett. Appl. Microbiol.* **2013**, *57*, 115–121. [CrossRef] [PubMed]

23. Merín, M.G.; de Ambrosini, V.I.M. Highly cold-active pectinases under wine-like conditions from non-*Saccharomyces* yeasts for enzymatic production during winemaking. *Lett. Appl. Microbiol.* **2015**, *60*, 467–474. [CrossRef] [PubMed]

24. Gallander, J.F.; Peng, A.C. Lipid and fatty acid composition of different wine grapes. *Am. J. Enol. Vitic.* **1980**, *31*, 24–27.

25. Pueyo, E.; Martinez-Rodriquez, A.; Polo, M.C.; Santa-Maria, G.; Bartomé, B. Release of lipids during yeast autolysis in model wine. *J. Agric. Food Chem.* **2000**, *48*, 116–122. [CrossRef] [PubMed]

26. Fragopoulou, E.; Antonopoulou, S.; Demopoulos, C.A. Biologically active lipids with antiatherogenic properties from white wine and must. *J. Agric. Food Chem.* **2002**, *50*, 2684–2694. [CrossRef] [PubMed]

27. Vakhlu, J.; Kour, A. Yeast lipases: Enzyme purification, biochemical properties and gene cloning. *Electron. J. Biotechnol.* **2006**, *9*. [CrossRef]

28. Matthews, A.; Grbin, P.R.; Jiranek, V. A survey of lactic acid bacteria for enzymes of interest in oenology. *Aust. J. Grape Wine Res.* **20026**, *12*, 235–244. [CrossRef]

29. Madrigal, T.; Maicas, S.; Tolosa, J.J.M. Glucose and ethanol tolerant enzymes produced by *Pichia* (*Wickerhamomyces*) isolates from enological ecosystems. *Am. J. Enol. Vitic.* **2013**, *64*, 126–133. [CrossRef]

30. Molnárova, J.; Vadkertiová, R.; Stratilová, E. Extracellular enzymatic activities and physiological profiles of yeasts colonizing fruit trees. *J. Basic Microbiol.* **2014**, *51*, S74–S84. [CrossRef] [PubMed]

31. Dimopoulou, M.; Lonvauf-Funel, A.; Dols-Lafargue, M. Polysaccharide production by grapes must and wine microorganisms. In *Biology of Microorganisms on Grapes, in Must and in Wine*, 2nd ed.; König, H., Unden, G., Fröhlich, J., Eds.; Springer International Publishing: Cham, Switzerland, 2017; pp. 241–258.

32. Pozo-Bayón, M.A.; Monagas, M.; Bartolomé, B.; Moreno-Arribas, M.V. Wine features related to safety and consumer health: An integrated perspective. *Crit. Rev. Food Sci. Nutr.* **2012**, *52*, 31–54. [CrossRef] [PubMed]

33. Campos, F.M.; Couto, J.A.; Hogg, T. Utilisation of natural and by-products to improve wine safety. In *Wine Safety, Consumer Preference, and Human Health*; Morena-Arribas, M.V., Bartolomé Sualdea, B., Eds.; Springer International Publishing: Cham, Switzerland, 2016; pp. 27–49.

34. Marchal, R.; Jeandet, P. Use of enological additives for colloids and tartrate salt stabilization in white wines and for improvement of sparkling wine foaming properties. In *Wine Chemistry and Biochemistry*; Morena-Arribas, M.V., Polo, M.C., Eds.; Springer: New York, NY, USA, 2010; pp. 127–158.

35. Venturi, F.; Andrich, G.; Quartacci, M.F.; Sanmartin, C.; Andrich, L.; Zinnai, A. A kinetic method to identify the optimum temperature for glucanase activity. *S. Afr. J. Enol. Vitic.* **2013**, *34*, 281–286. [CrossRef]

36. Styger, G.; Prior, B.; Bauer, F.F. Wine flavor and aroma. *J. Ind. Microbiol.* **2011**, *38*, 1145–1159. [CrossRef] [PubMed]

37. Ugliano, M.; Henschke, P.A. Yeasts and wine flavour. In *Wine Chemistry and Biochemistry*; Morena-Arribas, M.V., Polo, M.V., Eds.; Springer: New York, NY, USA, 2010; pp. 313–392.

38. Hjelmeland, A.K.; Ebeler, S.E. Glycosidically bound volatile aroma compounds in grapes and wine: A review. *Am. J. Enol. Vitic.* **2015**, *66*, 1–10. [CrossRef]

39. Mateo, J.J.; DiStefano, R. Description of the beta-glucosidase activity of wine yeasts. *Food Microbiol.* **1997**, *14*, 583–591. [CrossRef]

40. Maicas, S.; Mateo, J.J. Microbial glycosidases for wine production. *Beverages* **2016**, *2*, 20. [CrossRef]

41. Perez-Martin, F.; Sesena, S.; Miguel Izquierdo, P.; Martin, R.; Llanos Palop, M. Screening for glycosidase activities of lactic acid bacteria as a biotechnological tool in oenology. *World J. Microbiol. Biotechnol.* **2012**, *28*, 1423–1432. [CrossRef] [PubMed]

42. Michlmayer, H.; Nauer, S.; Brandes, W.; Schumann, C.; Kulbe, K.D.; del Hierro, A.M.; Eder, R. Release of wine monoterpenes from natural precursors by glycosidases from *Oenococcus oeni*. *Food Chem.* **2012**, *135*, 80–87. [CrossRef]

43. Iranzo, J.F.U.; Pérez, A.I.B.; Cañas, P.M.I. Study of oenological characteristics and enzymatic activities of wine yeasts. *Food Microbiol.* **1998**, *15*, 399–406. [CrossRef]

44. Zoecklein, B.; Marcy, J.; Williams, S.; Jasinski, Y. Effect of native yeasts and selected strains of *Saccharomyces cerevisiae* on glycosyl glucose, potential volatile terpenes, and selected aglycons of white Riesling (*Vitis vinfera* L.) wines. *J. Food Comp. Anal.* **1997**, *10*, 55–65. [CrossRef]

45. Wang, Y.; Zhang, C.; Xu, Y.; Li, J. Evaluating potential applications of indigenous yeasts and their β-glucosidases. *J. Inst. Brew.* **2015**, *121*, 642–648. [CrossRef]

46. Delcroix, A.; Gunata, Z.; Sapis, L.C.; Salmon, J.M.; Baynone, C. Glycosidase activities of three enological yeast strains during winemaking: Effect on the terpenol content of Muscat wine. *Am. J. Enol. Vitic.* **1994**, *45*, 291–296.

47. Sabel, A.; Martens, S.; Petri, A.; König, H.; Claus, H. *Wickerhamomyces anomalus* AS1: A new strain with potential to improve wine aroma. *Anal. Microbiol.* **2014**, *64*, 483–491. [CrossRef]

48. Rodríguez, M.E.; Lopes, C.A.; van Broock, M.; Valles, S.; Ramón, D.; Caballero, A.C. Screening and typing of Patagonian wine yeasts for glycosidase activities. *J. Appl. Microbiol.* **2004**, *96*, 84–95. [CrossRef] [PubMed]

49. Gonzales-Pombo, O.; Farina, L.; Carreau, F.; Batista-Viera, F.; Brena, B.M. A novel extracellular beta-glucosidase from *Issatschenkia terricola*: Immobilization and application for aroma enhancement of white Muscat wine. *Process Biochem.* **2011**, *46*, 385–389. [CrossRef]

50. Hu, K.; Zhu, X.L.; Mu, H.; Ma, Y.; Ullah, N.; Tao, Y.S. A novel extracellular glycosidase activity from *Rhodotorula mucilaginosa*: Its application potential in wine aroma enhancement. *Lett. Appl. Microbiol.* **2016**, *62*, 169–176. [CrossRef] [PubMed]

51. López, M.C.; Mateo, J.J.; Maicas, S. Screening of β-glucosidase and β-xylosidase activities in four non-*Saccharomyces* yeast isolates. *J. Food Sci.* **2015**, *80*, C1696–C1704. [CrossRef] [PubMed]

52. Schwentke, J.; Sabel, A.; Petri, A.; König, H.; Claus, H. The wine yeast *Wickerhamomyces anomalus* AS1 secretes a multifunctional exo-β-1,3 glucanase with implications for winemaking. *Yeast* **2014**, *31*, 349–359. [CrossRef] [PubMed]

53. El Rayess, Y. *Wine: Phenolic Composition, Classification and Health Benefits*; Nova Publishers: New York, NY, USA, 2014; ISBN 978-1-63321-048-6.

54. Padilla, B.; Gil, J.V.; Manzanares, P. Past and future of Non-*Saccharomyces* yeasts: From spoilage microorganisms to biotechnological tools for improving wine aroma complexity. *Front. Microbiol.* **2016**, 7, 111. [CrossRef] [PubMed]

55. Saerens, S.M.G.; Delvaux, F.; Verstrepen, K.J.; van Dijck, P.; Thevelein, J.M.; Delvaux, F.R. Parameters affecting ethyl ester production by *Saccharomyces cerevisiae* during fermentation. *Appl. Environ. Microbiol.* **2008**, 74, 454–461. [CrossRef] [PubMed]

56. Maicas, S.; Gil, J.V.; Pardo, I.; Ferrer, S. Improvement of volatile composition of wines by controlled addition of malolactic bacteria. *Food Res. Int.* **1999**, 32, 491–496. [CrossRef]

57. Matthews, A.; Grbin, P.R.; Jiranek, V. Biochemical characterization of the esterase activities of wine lacti acid bacteria. *Appl. Microbiol. Biotechnol.* **2007**, 77, 329–337. [CrossRef] [PubMed]

58. Fia, G.; Oliver, V.; Cavaglioni, A.; Canuti, V.; Zanoni, B. Side activities of commercial enzyme preparations and their influence on hydroxycinnamic acids, volatile compounds and nitrogenous components of white wine. *Aust. J. Grape Wine Res.* **2016**, 22, 366–375. [CrossRef]

59. Van Sluyter, S.C.; McRae, J.M.; Falconer, R.J.; Smith, P.A.; Bacic, A.; Waters, E.J.; Marangon, M. Wine protein haze: Mechanisms of formation and advances in prevention. *J. Agric. Food Chem.* **2015**, 63, 4020–4030. [CrossRef] [PubMed]

60. Weber, P.; Kratzin, H.; Brockow, K.; Ring, J.; Steinhart, H.; Paschke, A. Lysozyme in wine: A risk evaluation for consumers allergic to hen's egg. *Mol. Nutr. Food Res.* **2009**, 53, 1469–1477. [CrossRef] [PubMed]

61. Peñas, E.; di Lorenzo, C.; Uberti, F. Allergenic proteins in enology: A review on technological applications and safety aspects. *Molecules* **2015**, 20, 13144–13174. [CrossRef] [PubMed]

62. Rizzi, C.; Mainente, F.; Pasini, G.; Simonato, B. Hidden exogenous proteins in wine: Problems, methods of detection and related legislation—A review. *Czech J. Food Sci.* **2016**, 34, 93–104. [CrossRef]

63. Selitrennikoff, C.P. Antifungal proteins. *Appl. Environ. Microbiol.* **2001**, 67, 2883–2894. [CrossRef] [PubMed]

64. Jaeckels, N.; Tenzer, S.; Rosch, A.; Scholten, G.; Decker, H.; Fronk, P. β-Glucosidase removal due to bentonite fining during wine making. *Eur. Food Res. Technol.* **2015**, 241, 253–262. [CrossRef]

65. Esti, M.; Benucci, I.; Lombardelli, C.; Liburdi, K.; Garzillo, A.M.V. Papain from papaya (*Carica papaya* L.) fruit and latex: Preliminary characterization in alcoholic-acidic buffer for wine application. *Food Bioprod. Process.* **2013**, 91, 595–598. [CrossRef]

66. Benucci, I.; Esti, M.; Liburdi, K. Effect of free and immobilised stem bromelain on protein haze in white wine. *Aust. J. Grape Wine Res.* **2014**, 20, 347–352. [CrossRef]

67. Manrangon, M.; van Sluyter, S.C.; Robinson, E.M.C.; Muhlack, R.A.; Holt, H.E.; Haynes, P.A.; Godden, P.W.; Smith, P.A.; Waters, E.J. Degradation of white wine haze proteins by Aspergillopepsin I and II during flash pasteurization. *Food Chem.* **2012**, 135, 1157–1165. [CrossRef] [PubMed]

68. Van Sluyter, S.C.; Warnock, N.I.; Schmidt, S.; Anderson, P.; van Kan, J.A.L.; Bacic, A.; Waters, E.J. Aspartic acid protease from *Botrytis cinerea* removes haze-forming proteins during white winemaking. *J. Agric. Food Chem.* **2013**, 61, 9705–9711. [CrossRef] [PubMed]

69. Dizy, M.; Bisson, L.F. Proteolytic activity of yeast strains during grape juice fermentation. *Am. J. Enol. Vitic.* **2000**, 51, 155–167.

70. Younes, B.; Cilindre, C.; Villaume, S.; Parmentier, M.; Jeandet, P.; Vasserot, Y. Evidence for an extracellular and proteolytic activity secreted by living cells of *Saccharomyces cerevisiae* PIR1: Impact on grape proteins. *J. Agric. Food Chem.* **2011**, 59, 6239–6246. [CrossRef] [PubMed]

71. Younes, B.; Cilindre, C.; Jeandet, P.; Vasserot, Y. Enzymatic hydrolysis of thermos-sensitive grape proteins by a yeast protease as revealed by a proteomic approach. *Food Res. Int.* **2013**, 54, 1298–1301. [CrossRef]

72. Song, L.; Chen, Y.; Du, Y.; Wang, X.; Guo, X.; Dong, J.; Xiao, D. *Saccharomyces cerevisiae* proteinase A excretion and wine making. *World J. Microbiol. Biotechnol.* **2017**, 33, 210. [CrossRef] [PubMed]

73. Kang, H.A.; Choi, E.S.; Hong, W.K.; Kim, J.Y.; Ko, S.M.; Sohn, J.H.; Rhee, S.K. Proteolytic stability of recombinant human serum albumin secreted in the yeast *Saccharomyces cerevisiae*. *Appl. Microbiol. Biotechnol.* **2000**, 53, 575–582. [CrossRef] [PubMed]

74. Schlander, M.; Distler, U.; Tenzer, S.; Thines, E.; Claus, H. Purification and properties of yeast proteases secreted by *Wickerhamomyces anomalus* 227 and *Metschnikowia pulcherrima* 446 during growth in a white grape juice. *Fermentation* **2017**, 3, 2. [CrossRef]

75. Theron, L.W.; Bely, M.; Divol, B. Characterisation of the enzymatic properties of MpAPr1, an aspartic protease secreted by the wine yeast *Metschikowia pulcherrima*. *J. Sci. Food Agric.* **2017**, *97*, 3584–3593. [CrossRef] [PubMed]

76. Theron, L.W.; Bely, M.; Divol, B. Monitoring the impact of an aspartic protease (MpApr1) on grape proteins and wine properties. *Appl. Microbiol. Biotechnol.* **2018**, *102*, 5173–5183. [CrossRef] [PubMed]

77. Folio, P.; Ritt, J.F.; Alexandre, H.; Remize, F. Characterization of EprA, a major extracellular protein of *Oenococcus oeni* with protease activity. *Int. J. Food Microbiol.* **2008**, *127*, 26–31. [CrossRef] [PubMed]

78. Reid, V.J.; Theron, L.W.; du Toit, M.; Divol, B. Identification and partial characterization of extracellular aspartic protease genes from *Metschnikowia pulcherrima* IWBT Y1123 and *Candida apicola* IWBT Y1384. *Appl. Environ. Microbiol.* **2012**, *78*, 6838–6849. [CrossRef] [PubMed]

79. Mateo, J.J.; Maicas, S.; Thießen, C. Biotechnological characterisation of extracellular proteases produced by enological *Hanseniaspora* isolates. *Int. J. Food Sci. Technol.* **2015**, *50*, 218–225. [CrossRef]

80. Lagace, L.S.; Bisson, L.F. Survey of yeast acid proteases for effectiveness of wine haze reduction. *Am. J. Enol. Vitic.* **1990**, *41*, 147–155.

81. Chasseriaud, L.; Miot-Sertier, C.; Coulon, J.; Iturmendi, N.; Moine, V.; Albertin, W.; Bely, M. A new method for monitoring the extracellular proteolytic activity of wine yeasts during alcoholic fermentation of grape must. *J. Microbiol. Methods* **2015**, *119*, 176–179. [CrossRef] [PubMed]

82. Claus, H.; Sabel, A.; König, H. Wine phenols and laccase: An ambivalent relationship. In *Wine, Phenolic Composition, Classification and Health Benefits*; Rayess, Y.E., Ed.; Nova Publishers: New York, NY, USA, 2014; pp. 155–185.

83. Claus, H.; Decker, H. Bacterial tyrosinases. *Syst. Appl. Microbiol.* **2006**, *29*, 3–14. [CrossRef] [PubMed]

84. Fronk, P.; Hartmann, H.; Bauer, M.; Solem, E.; Jaenicke, E.; Tenzer, S.; Decker, H. Polyphenoloxidase from Riesling and Dornfelder wine grapes (*Vitis vinfera*) is a tyrosinase. *Food Chem.* **2015**, *183*, 49–57. [CrossRef] [PubMed]

85. Riebel, M.; Sabel, A.; Claus, H.; Fronk, P.; Xia, N.; Li, H.; König, H.; Decker, H. Influence of laccase and tyrosinase on the antioxidant capacity of selected phenolic compounds on human cell lines. *Molecules* **2015**, *20*, 17194–17207. [CrossRef] [PubMed]

86. Riebel, M.; Sabel, A.; Claus, H.; Xia, N.; Li, H.; König, H.; Decker, H.; Fronk, P. Antioxidant capacity of phenolic compounds on human cell lines as affected by grape-tyrosinase and *Botrytis*-laccase oxidation. *Food Chem.* **2017**, *229*, 779–789. [CrossRef] [PubMed]

87. Sabel, A.; Bredefeld, S.; Schlander, M.; Claus, H. Wine phenolic compounds: Antimicrobial properties against yeasts, lactic acid and acetic acid bacteria. *Beverages* **2017**, *3*, 29. [CrossRef]

88. Zinnai, A.; Venturi, F.; Sanmartin, C.; Quartacci, M.F.; Andrich, G. Chemical and laccase catalysed oxidation of gallic acid: Determination of kinetic parameters. *Res. J. Biotechnol.* **2013**, *6*, 62–65.

89. Servili, M.; de Stefano, G.; Piacquadio, P.; Sciancalepore, V. A novel method for removing phenols from grape must. *Am. J. Enol. Vitic.* **2000**, *51*, 357–361.

90. Minussi, R.C.; Rossi, M.; Bolgna, L.; Rotilio, D.; Pastore, G.M.; Durán, N. Phenols removal in musts; strategy for wine stabilization by laccase. *J. Mol. Catal. B Enzym.* **2007**, *45*, 102–107. [CrossRef]

91. Maier, G.; Dietrich, H.; Wucherpfennig, K. Winemaking without SO_2-with the aid of enzymes? *Weinwirtschaft-Technik* **1990**, *126*, 18–22.

92. Brenna, O.; Bianchi, E. Immobilized laccase for phenolic removal in must and wine. *Biotechnol. Lett.* **1994**, *16*, 35–40. [CrossRef]

93. Lettera, V.; Pezzella, C.; Cicatiello, P.; Piscitelli, A.; Giacobelli, V.G.; Galano, E.; Amoresano, A.; Sannia, G. Efficient immobilization of a fungal laccase and its exploitation in fruit juice clarification. *Food Chem.* **2016**, *196*, 1272–1278. [CrossRef] [PubMed]

94. Lustrato, G.; De Leonardis, A.; Macciola, V.; Ranalli, G. Preliminary lab scale of advanced techniques as new tools to reduce ethylphenols content in synthetic wine. *Agro Food Ind. Hi-Tech* **2015**, *26*, 51–54.

95. Smit, A.A.; du Toit, W.J.; du Toit, M. Biogenic amines in wine: Understanding the headache. *S. Afr. J. Enol. Vitic.* **2008**, *29*, 109–127. [CrossRef]

96. Guo, Y.Y.; Yang, Y.P.; Peng, Q.; Han, Y. Biogenic amines in wine: A review. *Int. J. Food Sci. Technol.* **2015**, *50*, 1523–1532. [CrossRef]

97. Preti, R.; Vieri, S.; Vinci, G. Biogenic amine profiles and antioxidant properties of Italian red wines from different price categories. *J. Food Comp. Anal.* **2016**, *46*, 7–14. [CrossRef]

98. Sebastian, P.; Herr, P.; Fischer, U.; König, H. Molecular identification of lactic acid bacteria occurring in must and wine. *S. Afr. J. Enol. Vitic.* **2011**, *32*, 300–309. [CrossRef]
99. Christ, E.; König, H.; Pfeiffer, P. Bacterial formation of biogenic amines in grape juice: Influence of culture conditions. *Deutsche Lebensmittel-Rundschau* **2012**, *108*, 73–78.
100. Henríquez-Aedo, K.; Durán, D.; Garcia, A.; Hengst, M.B.; Aranda, M. Identification of biogenic amines-producing lactic acid bacteria isolated from spontaneous malolactic fermentation of Chilean red wines. *LWT Food Sci. Technol.* **2016**, *68*, 183–189. [CrossRef]
101. Moreno-Arribas, M.V.; Polo, M.C. (Eds.) Amino acids and biogenic amines. In *Wine Chemistry and Biochemistry*; Springer: New York, NY, USA, 2010; pp. 163–189.
102. Kushnereva, E.V. Formation of biogenic amines in wine production. *Appl. Biochem. Microbiol.* **2015**, *51*, 108–112. [CrossRef]
103. Yagodina, O.V.; Nikol'skaya, E.B.; Khovanskikh, A.E.; Kormilitsyn, B.N. Amine oxidases of microorganisms. *J. Evol. Biochem. Physiol.* **2002**, *38*, 251–258. [CrossRef]
104. Klinman, J.P. The multi-functional topa-quinone copper amine oxidases. *Biochim. Biophys. Acta* **2003**, *1647*, 131–137. [CrossRef]
105. Corpillo, D.; Valetti, F.; Giuffrida, M.G.; Conti, A.; Rossi, A.; Finazzi-Agrò, A.; Giunta, C. Induction and characterization of a novel amine oxidase from the yeast *Kluyveromyces marxianus*. *Yeast* **2003**, *20*, 369–379. [CrossRef] [PubMed]
106. Bäumlisberger, M.; Moellecken, U.; König, H.; Claus, H. The potential of the yeast *Debaryomyces hansenii* H525 to degrade biogenic amines in food. *Microorganisms* **2015**, *3*, 839–850. [CrossRef] [PubMed]
107. Callejón, S.; Sendra, R.; Ferrer, S.; Pardo, I. Identification of a novel enzymatic activity from lactic acid bacteria able to degrade biogenic amines in wine. *Appl. Microbiol. Biotechnol.* **2014**, *98*, 185–198. [CrossRef] [PubMed]
108. Callejón, S.; Sendra, R.; Ferrer, S.; Pardo, I. Cloning and characterization of a new laccase from *Lactobacillus plantarum* J16 CECT 8944 catalyzing biogenic amine degradation. *Appl. Microbiol. Biotechnol.* **2016**, *100*, 3113–3124. [CrossRef] [PubMed]
109. Lonvaud-Funel, A. Undesirable compounds and spoilage microorganisms in wine. In *Wine Safety, Consumer Preference, and Human Health*; Morena-Arribas, M.V., Bartolomé Sualdea, B., Eds.; Springer International Publishing: Cham, Switzerland, 2016; pp. 3–26.
110. König, H.; Fröhlich, J. Lactic acid bacteria. In *Biology of Microorganisms on Grapes, in Must and in Wine*, 2nd ed.; Springer International Publishing: Cham, Switzerland, 2017.
111. Liburdi, K.; Benucci, I.; Esti, M. Lysozyme in wine: An overview of current and future applications. *Compr. Rev. Food Sci. Food Saf.* **2014**, *13*, 1062–1073. [CrossRef]
112. Mojsov, K.; Petreska, M.; Ziberoski, J. Risks of microbial spoilage in wine. In Proceedings of the EHEDG World Congress on Hygienic Engineering & Design, Ohrid, Macedonia, 22–24 September 2011.
113. Blättel, V.; Wirth, K.; Claus, H.; Schlott, P.; Pfeiffer, P.; König, H. A lytic enzyme cocktail from *Streptomyces* sp. B578 for the control of lactic and acid bacteria in wine. *Appl. Microbiol. Biotechnol.* **2009**, *83*, 839–848. [CrossRef] [PubMed]
114. Sommer, S.; Wegmann-Herr, P.; Wacker, M.; Fischer, U. Influence of lysozyme addition on hydroxycinnamic acids and volatile phenols during wine fermentation. *Fermentation* **2018**, *4*, 5. [CrossRef]
115. Gonzales, R.; Tronchoni, J.; Quirós, M.; Morales, P. Genetic improvement and genetically modified microorganisms. In *Wine Safety, Consumer Preference, and Human Heath*; Morena-Arribas, M.V., Bartolomé Sualdea, B., Eds.; Springer International Publishing: Cham, Switzerland, 2016; pp. 71–96.
116. Pretorius, I.S. Tailoring wine yeast for the new millenium: Novel approaches to the ancient art of winemaking. *Yeast* **2000**, *16*, 675–729. [CrossRef]
117. Pretorius, I.S. Conducting wine symphonics with the aid of yeast genomics. *Beverages* **2016**, *2*, 36. [CrossRef]

fermentation

MDPI

Review

A Future Place for *Saccharomyces* Mixtures and Hybrids in Wine Making

Helmut König * and Harald Claus

Institute for Molecular Physiology, Johannes Gutenberg-University, Johann-Joachim-Becher-Weg 15, 55128 Mainz, Germany; hclaus@uni-mainz.de
* Correspondence: Helmut.Koenig.Ingelheim@t-online.de

Received: 26 June 2018; Accepted: 15 August 2018; Published: 18 August 2018

Abstract: Each year, winemakers can face sluggish or stuck fermentations during wine making, especially when a spontaneous fermentation is performed, even if strains of the classical wine yeast *Saccharomyces cerevisiae* are applied. Problems are inevitable when low ammonium concentrations (<160 mg L^{-1} grape must) or an excess of fructose compared to glucose are observed during grape must fermentation. *S. cerevisiae* strains cannot use all kinds of amino acids as the sole nitrogen source but usually need free ammonium (optimal concentration: 600 mg L^{-1} grape must). It preferably consumes glucose, leading often to an excess of fructose in the fermenting must, which contains glucose and fructose in an equal ratio at the beginning of fermentation. Yeast hybrids have been isolated from wines several times and different strains are already commercially available. The united properties of the parent strains can provide advantages under sophisticated fermentation conditions. However, the involvement of a hybrid yeast for the rectification of fermentation disorders in spontaneous fermentations has only been described recently in the literature. Recent investigations have provided convincing evidence that fermentation problems can be overcome when must fermentations are successively performed with *Saccharomyces bayanus* strain HL 77 and the triple hybrid *S. cerevisiae* × *Saccharomyces kudriavzevii* × *S. bayanus* strain HL 78. The triple hybrid strain HL 78 uses amino acids as a nitrogen source in the absence of ammonium and it also exhibits a fructophilic character with an enhanced uptake of fructose in comparison to glucose. The application of genetically modified yeast strains is not allowed for starter cultures in wine making, but the usage of yeast mixtures and hybrid strains could be a promising tool for winemakers to solve fermentation problems during spontaneous fermentation or for the creation of novel wine types with desired sensory characteristics under more challenging conditions, especially when the composition of the must components is not optimal because of, e.g., critical climatic or soil conditions.

Keywords: *Saccharomyces*; yeast hybrids; yeast mixtures; spontaneous fermentation; stuck and sluggish fermentation

1. Introduction

In publications about the history of wine making, McGovern [1,2] and Kupfer [3,4] provided convincing indications that cultures of vines and wine making were established between 6000 and 8000 BC in regions between the Black Sea and the Caspian Sea and also along the later Silk Road all the way to China. Successively, viticulture spread via different countries of Asia Minor and northern Africa. It arrived in about 1000 BC in southern European countries such as Italy, France, and Spain. In the end, wine production was well established in more northern and eastern parts of Europe around 1000 AD. Today, the most common vine variety around the earth is *Vitis vinifera* L. subsp. *vinifera*. This variety arose from the Eurasian wild form of *Vitis vinifera* L. subsp. *sylvestris*.

Wine production has a long tradition and has been a part of human culture for thousands of years. Despite this fact, deeper insights into the microbiology and biochemistry of the conversion of the sugars in grape must to ethanol and into other accompanying biochemical reactions were obtained only since the 19th century. In the times before, the art of wine making had been further developed mostly empirically from generation to generation. Fundamental scientific investigations about biochemical transformations during wine making started not before the end of the 18th century and in the course of the 19th century [5–9]. Although, *Saccharomyces cerevisiae* likely played the most important role in wine fermentation from the beginning of viticulture [10], it was not before 1883 that the first pure yeast culture was obtained by Emil Christian Hansen. Originally, these isolates were used for beer production, while around the year 1890, Hermann Mueller-Thurgau also introduced yeast starter cultures to wine making. Only since the 1930s have commercial liquid cultures of yeasts been available as starter cultures for the inoculation of must, which were commonly used after the Second World War.

The variety of microbes growing in fermenting must is limited to three groups of ethanol- and acid-tolerant microorganisms, namely, yeasts, lactic acid bacteria, and acetic acid bacteria [11]. From grapes, must, and wine, more than 100 yeast species belonging to 49 genera have been isolated and characterized [12–15]. The classical wine yeast *S. cerevisiae* and the so-called wild yeasts (non-Saccharomycetes) are involved in the conversion of must into wine. The varieties and succession of yeast species during the fermentation of a certain must sample have a significant impact on the specific sensory profile of the produced wine. Compared to the known wine-related yeasts species, the variety of bacterial species is lower. Around 25 species of wine-related lactic acid bacteria have been obtained in pure culture. They belong to the genera *Lactobacillus*, *Leuconostoc*, *Oenococcus*, *Pediococcus*, and *Weissella* [16]. In addition, 23 acetic acid bacteria have been detected on grapes, in must, and in wine, which belong to the genera *Acetobacter*, *Amayamea*, *Asaia*, *Gluconacetobacter*, *Gluconobacter*, *Komagateibacter*, and *Kozakia* [17]. Molecular biology methods and next-generation sequencing approaches have been applied to examine and quantify the diversity and genetic variations as well as sources and roles of wine-related microorganisms. This knowledge is also helpful for the development of novel starter cultures [18–21].

Up to now, a relatively broad knowledge of the diversity, succession, and physiological and biochemical activities of wine-related microorganisms has been acquired. Mainly, the yeast species *S. cerevisiae* and the two bacterial species *Oenococcus oeni* or *Lactobacillus plantarum* are commercially available and applied for alcoholic and malolactic fermentation, respectively. Despite the deeper microbiological and biochemical knowledge of the backgrounds of wine making, sluggish or stuck fermentations cause significant financial losses for winemakers each year. These unwanted observations stimulate investigations for more improved microbiological strains and novel procedures to circumvent the observed fermentation obstacles.

2. Today's Principal Procedures and Obstacles of Wine Making

In the last decades, the risk of sluggish or stuck fermentations has been significantly reduced by the commercial availability and application of selected strains of the classical wine yeast *S. cerevisiae*, especially when about 10^5 cells/mL are added by winemakers to start controlled fermentation. Today, a great variety of *S. cerevisiae* strains is offered by different companies for the production of wines with different sensory profiles. Because of the high titer of the starter yeast cells, wild yeasts have difficulties developing and fermentation can be carried out relatively reproducibly by starter cultures. While this procedure greatly reduces the risk of fermentation problems, the sensory profiles compared to a spontaneous fermentation are restricted and depend on the starter cultures used.

On the other hand, monitored fermentation is started spontaneously and selected yeast cultures are only added when fermentation problems are observed. Satisfying results can be obtained with optimized yeast strains or yeast mixtures. For this procedure, we isolated and selected yeast strains of *S. cerevisiae* from fermenting must in a certain vineyard in previous years. Harvested cells from grown fermenter cultures were then only added to sluggishly fermenting must in the same vineyard in order to continue and finish the fermentation. Compared to a solely spontaneous fermentation, by this method, relatively complex wines were also produced which met the special sensory requirements of the winemaker quite well (unpublished special service for certain wineries, provided by the Institute of Microbiology and Wine Research of the Johannes Gutenberg-University in Mainz, Germany).

For millennia, wines have been made by spontaneous fermentation, which is therefore the earliest form of must fermentation. In this case, the yeast strains present in the cellar or on the grapes enter the must and start the fermentation. At the beginning of the fermentation, the classical wine yeast *S. cerevisiae* is present only at a relatively low cell number. The first stage of a spontaneous fermentation is usually dominated by a mixture of some species of so-called wild yeasts. The indigenous wild yeasts, the classic wine yeast *S. cerevisiae*, and the local lactobacilli as well as acetic acid bacteria are involved in the microbial conversion of grape must into wine. Usually, when the ethanol concentration reaches about 4% to 7% (*v/v*), *S. cerevisiae* can overgrow the wild yeasts and also most of the bacterial strains in the fermenting must. The corresponding wines are often more complex and are more likely to meet the expectations of a particular terroir. The risk of fermentation problems, however, is obviously increased compared to controlled and monitored fermentations.

Despite the observed increased reliability of must fermentation by adding starter cultures of commercial yeast strains after grape pressing, winemakers, especially of the upper-quality segment, often favor spontaneous fermentation in order to produce more complex wines with a characteristic sensory profile distinctive for a certain winery or terroir. Of course, the sensory profile relies not only on multifactorial environmental and biological features, such as the grape variety and grape quality, the terroir (soil and climate), the conditions in the wine cellar, and the fermentation management, but also, without doubt, to a greater part on the added or indigenous microbiota. In the case of spontaneous fermentation, the bacteria and yeast composition in the fermenting must depends on the microorganisms on the grapes and in the cellar. However, it should be remembered that the risk of fermentation problems in the case of spontaneous fermentation is increased. Some reasons are well known to be responsible for fermentation problems. These include (a) heavily infected grapes, (b) low and fluctuating temperatures, (c) toxic and fungicidal compounds, (d) killer toxins, (e) ratio of glucose to fructose below 1:10, (f) deficiencies of nutrients such as vitamins or trace elements, (g) ammonium concentration below 120 mg/L, (h) pH values below 3.0, and (i) elevated polyphenol concentrations [22–25].

Ordinary attempts to overcome the observed fermentation problems are (a) adjustment of the temperature to 20 °C, (b) addition of yeast nutrients (diammonium hydrogen phosphate), (c) increase of the pH value, and (d) a reinoculation with yeast starter cultures possessing a high ethanol tolerance (e.g. sparkling wine yeasts). However, these measures frequently lead to a change in the initially targeted sensory profile, which is hardly compatible with the conceptions of winemakers in the upper-quality segments and their very special sensory expectations.

3. Suggested Efforts for More Sophisticated Wine Production

3.1. Application of Wild Yeasts and Yeast Mixtures

In general, wild yeasts are suggested to be suitable tools for wine tailoring. They can have an influence on low sulfite formation, reduction of copper content, reduction of ochratoxin A, reduced production of ethyl carbamate, low biogenic amine formation, reducing volatile acidity, alcohol reduction, modulation of acidity, increased glycerol content, modulation of aroma profiles, enhancing varietal aromas, mannoprotein release, and control of spoilage microflora [12,26–28].

When a spontaneous fermentation is started, the wild yeasts (non-Saccharomycetes) dominate in number compared to the classical wine yeast *S. cerevisiae* [12]. The growth of the non-Saccharomycetes is more sensitive to increasing sulfite and ethanol concentrations. Some of them, such as *Torulaspora delbrueckii* and *Lachancea* (*Kluyveromyces*) *thermotolerans*, require higher oxygen concentrations for optimal growth than *S. cerevisiae*. The reason is a different synthesis rate of unsaturated fatty acids under oxygen-limiting conditions [27,29–32]. Sometimes *S. cerevisiae* grows suboptimally in the presence of wild yeasts which already partly consume nutrients and growth-promoting substances such as sugars, trace elements, or vitamins. Non-Saccharomycetes can produce acetic acid, which is unfavorable for Saccharomycetes. Acetic acid can be sensed at a concentration of 0.6 g/L. The Old World and the New World wine regulations are quite different. Wine making is strictly ruled in the Old World (e.g., France: Appellation d'origine contrôlée (AOC), Italy: Denominazione di origine controllata (DOC), Spain: Denominación de Origen (DO), Portugal: Denominação de Origem Controlada (DOC), Germany: German Wine Law). The regulations are more relaxed and open for innovations in the New Word, but in some cases, more restrictive rules have been established (e.g., Vintner's Quality Alliance, VQA). In the European Economic Community (EEC) (renamed as European Community (EC) after formation of the European Union (EU) in 1993), values are 1.07 g/L for white wine and 1.20 g/L for red wine. The Common Agricultural Policy (CAP) of the EU includes the EU wine regulations. The member states produce about 65% of the global wine. The details of quality classifications are part of the national wine laws. According German law, the upper limits for white and red wines is 1.08 g/L and 1.20 g/L, respectively. For wines of individually selected overripe berries (Trockenbeerenauslese), 2.10 g/L is allowed. In France, wine is of commercial quality according to the appellation d'origine controlée (AOC system) if the acetic acid concentration does not exceed 1.1 g/L. Due to the Australia New Zealand Food Standards Code - Standard 4.5.1—Wine Production Requirements (Australia Only), wine, sparkling wine, and fortified wine must contain no more than 1.5 g/L of volatile acidity, excluding sulfur dioxide, expressed as acetic acid. According to the Code of Federal Regulations (CFR) of the Federal Government of the United States (CFR title 27/4.21), the maximum volatile acidity, calculated as acetic acid and exclusive of sulfur dioxide, is 0.14 gram per 100 mL (20 °C) for natural red wine and 0.12 gram per 100 mL (20 °C) for other grape wines.

The concentration of ethanol is reduced when this alcohol is partly oxidized and converted into the volatile compound acetic acid. Polysaccharide production by wild yeasts can lead to graisse at higher concentrations, while a positive mouth feeling is perceived at a certain concentration. Polysaccharides have an influence on the sensory properties of wines because of their interactions with wine and salivary proteins, tannins, tartrate, and aroma compounds [10]. Wines made with yeast strains that produce inherently higher levels of polysaccharides are naturally softer, have more body, and a better mouthfeel. Regular "batonnage" in this period can further stimulate the release of polysaccharides, which in turn will have a positive influence on the mouthfeel and body of the wine [33–35]. In addition, different fruit and unwanted aromas are caused by the formation of a large variety of esters.

When *L. thermotolerans* and *Candida zemplinina* grow in must, the glycerol content can be increased, which also has an influence on the mouth feeling. *C. zemplinina* can also lower the concentration of the acetic acid produced in must possessing a high sugar content. Members of the genera *Debaryomyces*, *Hansenula*, *Candida*, *Pichia*, and *Kloeckera* can produce aromatic and colored compounds (anthocyanins) from glycoconjugates. An inoculated mixture of *Debaryomyces pseudopolymorphus* and *S. cerevisiae* increased concentrations of terpenols such as citronell, nerol, and geraniol in Chardonnay wines [27]. After addition of *C. zemplinina* and *Pichia kluyveri* to a sample of a Sauvignon Blanc must, the concentration of sulfur-containing compounds, such as 3-mercaptohexan-1-ol (3MH) and 3-mercaptohexan-1-ol acetate (3MHA), was increased. Thioles contribute to the specific aroma of Sauvignon Blanc wines [27]. The yeast *Wickerhamomyces anomalus* can hydrolyze a series of glycosylated aroma precursors [36,37].

Wild yeasts are dominant in the first stage of the must fermentation. Members of the genera *Hanseniaspora*, *Rhodotorula*, *Pichia*, *Candida*, *Metschnikowia*, and *Cryptococcus* were often identified at the beginning of alcoholic fermentation [15,27]. Wild yeasts can be divided into different physiological groups. Species of *Pichia*, *Debaryomyces*, *Rhodotorula*, and *Candida* prefer aerobic growth. *Hanseniaspora uvarum* (perfect form: *Kloeckera apiculata*), *Hanseniaspora guilliermondii* (perfect form: *Kloeckera apiculata* var. *apis*), and *Hanseniaspora occidentalis* (perfect form: *Kloeckera javanica*) have a low fermentation activity, while strains of *Kluyveromyces marxianus*, *T. delbrueckii*, *Metschnikowia pulcherrima*, and *Zygosaccharomyces bailii* exhibit increased fermentation activities [27,38].

In order to realize certain desirable aromas in wine, fermentation can be started by the addition of individual selected wild yeast species and then continued with a starter culture of the classical wine yeast *S. cerevisiae* in order to complete the fermentation. Wild yeast cultures and also yeast mixtures are already commercially available. For instance, *Torulaspora delbrückii* strains are offered under the designation "Oenoferm wild & pure". Some commercial cultures contain, in addition to *S. cerevisiae* wild yeasts (20–40%), for example, *K. thermotolerans* or *T. delbrueckii*. A mixed culture consisting of *L. (K.) thermotolerans* (20%), *T. delbrueckii* (20%), and *S. cerevisiae* (60%) is also available. The two known starter cultures Sihaferm PireNature or Level 2 TD are made up of *T. delbrueckii* and *S. cerevisiae*. In this case the must fermentation is started with wild yeast and then continued and completed with *S. cerevisiae*. Viniflora® PRELUDE™ contains a *Torulaspora* starter culture which is added to the must at concentrations of 20 g/hL. The inoculated must is then kept at 7–10 °C for 4–7 days. At an ethanol concentration of 4–6 vol%, a *Saccharomyces* strain is added in the same amount after transfer of the fermenting must to a fermentation tank. The temperature is then increased. Finally, a surplus of malolactic acid can be biologically reduced with the lactic acid bacterium *O. oeni*.

3.2. Application of Hybrid Yeasts for Overcoming Fermentation Problems

During spontaneous fermentation, wine-related microbial species usually grow in succession. In the first stage of the fermentation, the so-called wild yeasts (non-Saccharomycetes) are multiplying. In the harvest years 2011 and 2012, the succession of the microorganisms in the course of the spontaneous fermentation of Riesling must in the winery Heymann-Löwenstein (lower Moselle, Germany) was investigated [39,40]. The wild yeasts in a wine cask without fermentation problems belonged to the genera/species *Candida pararuqosa*, *Saccharomycetes* sp./*Pichia membranifaciens*, *Saccharomycopsis crateagensis*, *Candida boidinii*, *Saccharomycetes* sp., *Aureobasidium* sp., *Metschnikowia* sp., *Metschnikowia chrysoperlae*, *Cryptococcus flavescens*, *C. zemplinina*, *P. kluyveri*, and *H. uvarum*. Interestingly, in some barrels, wild yeast species survived at elevated levels of ethanol. Living cells of *C. boidinii* were found until end of fermentation. The genus *Saccharomyces* contains nine species [41]. Unexpectedly, the fermentation was not initiated by the classical wine yeast *S. cerevisiae* but rather by *Saccharomyces bayanus* [39]. Approximately 4 weeks after an observed stuck fermentation, the alcoholic fermentation was completed by the triple hybrid *S. cerevisiae* × *Saccharomyces kudriavzevii* × *S. bayanus* strain HL78. This hybrid possessed genome sequences of the three mentioned *Saccharomyces* species. The triple hybrid yeast strain HL 78 was not added to the must but grew in the background during the fermentation in the must. Therefore, strain HL 78 must have been already present in low cell numbers after fermentation started. The classical wine yeast *S. cerevisiae* was not able to grow in the investigated must because the temperature in the wine cellar was between 12 and 14 °C and the temperature in the wine cask reached only about 16 °C at the most. The different yeast strains of *S. cerevisiae* require more than 140 mg nitrogen/L for optimal growth [38,39,42]. In the investigations presented here, the available ammonium nitrogen decreased after starting the fermentation from 120 to 40 mg/L in a relatively short time [39]. It is well known that at low ammonium concentrations, the sugar uptake activity also decreases in the case of *S. cerevisiae* [38,39]. In comparison to the classical wine yeast *S. cerevisiae*, *S. bayanus* can also grow better at low temperatures and low available ammonium concentrations. *S. bayanus* strain HL 77 and in particular the triple hybrid strain HL 78 are able to satisfy their nitrogen needs from amino acids or probably proteins, even at low concentrations and without free ammonium. As demonstrated by quantitative proteomics, higher protease activities were detected in the triple hybrid strain HL 78 compared to *S. cerevisiae* [40]. The triple hybrid HL 78 could uptake glucose and especially fructose at lower amino acid concentrations. In addition, the triple hybrid strain exhibits a fructophilic character, which is reflected in a higher uptake rate of radiolabeled fructose compared to glucose [43].

The hybrid *S. cerevisiae* × *S. kudriavzevii* was described by González et al. [44]. The combined characteristics of both parents may be advantageous under certain fermentation conditions. The hybrid is tolerant against high ethanol concentrations and high osmolarity, which is a characteristic feature of

S. cerevisiae. The tolerance against cool temperatures is a feature of *S. kudriavzevii*. Yeast hybrids have been isolated in Europe by several authors [45–47], although the strains of *S. kudriavzevii* known so far have been isolated from decaying leaves in Japan [48] and from oak bark samples in Portugal [49]. However, the involvement of a hybrid yeast for the elimination of fermentation disorders in spontaneous fermentations has not been mentioned in the literature so far.

4. Discussion

For a few thousand years, must has been fermented spontaneously without the knowledge of the specific physiological and biochemical activities of the involved microorganisms. Therefore, progress in the art of wine making was only gained empirically. In the second half of the last century, spontaneous fermentations were largely successively replaced by the application of selected starter cultures of *S. cerevisiae*, leading to regular use since the 80s of the last century by many winegrowers and cooperatives.

The available improved scientific and practical knowledge led to considerable progress in wine growing and vinification in the last decades. The application of more selected yeasts and methods enables a much more defined control of the fermentation process. The application of molecular biology identification methods and the sequence analysis of nucleic acids have shown that diverse yeast strains occur in the different wine-growing regions which enable the use of region-specific starter cultures after isolation of pure strains [50–53]. Probably occurring fermentation problems during a spontaneous fermentation can thus be remedied by the subsequent addition of terroir-specific yeast strains without the need to accept major changes in the desired flavor profile. Moreover, a partial imitation of spontaneous fermentation is now possible by the use of isolated and selected strains of wild yeasts.

Strategies for the targeted genetic modification of yeasts can in principle be worked out or have already been described [54,55] in order to produce yeast strains with certain desired properties. However, genetically modified yeasts are not authorized for wine making in Europe or Australia. The recognition of safe (GRAS) in the United States applies only for two strains (ML01 and 533EC). Yeast starter cultures can be furthermore improved by selecting certain strains by evolutionary in vitro adaptation or by the production of hybrids. The so-called "evolutionary in vitro adaptation" is performed by a slow change of the culture conditions during several months. A fructophilic yeast was obtained from a normal *S. cerevisiae* isolate by slowly shifting the glucose/fructose ratio towards fructose (Pfeiffer, P.; König, H. unpublished results).

Several yeast hybrid strains are already commercially available. Here, only some examples can be given: strain "Oenoferm® X-treme" is a GMO-free hybrid yeast obtained from the protoplast fusion of two different *S. cerevisiae* strains; strain "Cross Evolution" is a natural cross hybrid between *S. cerevisiae* yeasts; strain NT 202 is a product of the yeast hybridization program; strain S6U is a hybrid of *S. cerevisiae* × *S. bayanus*; and strain VIN7 is an allotriploid interspecific hybrid of a heterozygous diploid complement of *S. cerevisiae* chromosomes and a haploid *S. kudriavzevii* genomic contribution [56].

In the future, the art of wine making includes the simultaneous or sequential use of different strains of *S. cerevisiae* as in the past, but defined mixtures of different species of Saccharomycetes and non-Saccharomycetes will also be used more generally. This will enable quite different wine styles. The future belongs to well-trained creative winemakers who can handle the different protocols with varying compositions of starter cultures to stimulate the microbe orchestra to new sounds.

Conflicts of Interest: The authors declare no conflicts of interest.

References

1. McGovern, P.E. *Acient Wine. The Search for the Origins of Viniculture*; Princeton University Press: Princeton, NJ, USA, 2003.

2. McGovern, P.E.; Zhang, J.; Tang, J. Fermented beverages of pre- and proto-historic China. *Proc. Natl. Acad. Sci. USA* **2004**, *101*, 261–270. [CrossRef] [PubMed]
3. Kupfer, P. Weinstraße vor der Seidenstraße—Weinkulturen zwischen Georgien und China. In *Kulturgut Rebe und Wein*; König, H., Decker, H., Eds.; Springer Spektrum: Heidelberg, Germany, 2013; pp. 3–17.
4. Kupfer, P. Der älteste Wein der Menschheit in China: Jiahu und die Suche nach den Ursprüngen der eurasischen Weinkultur. In *Wo aber der Wein fehlt, stirbt der Reiz des Lebens*; Decker, H., König, H., Zwickel, W., Eds.; Nünnerich-Asmus Verlag: Mainz, Germany, 2015; pp. 12–25.
5. Gay-Lussac, J.L. Sur l'analyse de l'alcohol et de l'ether sulfurique et sur les produits de la fermentation. *Annales de Chimie et de Physique* **1815**, *95*, 311.
6. Schwann, T. Vorläufige Mitteilung, betreffend Versuche über die Weingährung und Fäulnis. *Annalen der Physik und Chemie* **1837**, *41*, 184–193. [CrossRef]
7. Kützing, F.T. Mikroskopische Untersuchungen über die Hefe und Esssigmutter, nebst mehreren anderen dazu gehörigen vegetabilischen Gebilden. *J. Prakt. Chem.* **1837**, *11*, 385–409. [CrossRef]
8. Cagniard-Latour, C.C. Mémoire sur la fermentation vineuse. *Ann. Chim. Phys.* **1838**, *68*, 206–222.
9. Pasteur, M.L. Animalcules infusoires vivant sans gaz oxygène libre et détermination des fermentation. *Comptes Rendus de l'Academie des Sciences* **1861**, *52*, 344–347.
10. Feldmann, H. *Yeast. Molecular and Cell Biology*; Wiley-Blackwell: Weinheim, Germany, 2010.
11. König, H.; Unden, G.; Fröhlich, J. (Eds.) Biology of Microorganisms on Grapes. In *Must and in Wine*, 2nd ed.; Springer: Heidelberg, Germany, 2017.
12. Bisson, L.F.; Joseph, C.M.L.; Domizio, P. Yeasts. In *Biology of Microorganisms on Grapes, in Must and in Wine*, 2nd ed.; König, H., Unden, G., Fröhlich, J., Eds.; Springer: Heidelberg, Germany, 2017; pp. 65–105.
13. Fugelsang, K.C.; Edwards, C.G. *Wine Microbiology. Practical Applications and Procedures*; Springer: Heidelberg, Germany, 2007.
14. Fleet, G.H.; Heard, G.M. Yeasts—Growth during fermentation. In *Wine Microbiology and Biotechnology*; Fleet, G.H., Ed.; Harwood Academic Publishers: Chur, Switzerland, 1993; pp. 27–54.
15. König, H.; Berkelmann-Löhnertz, B. Maintenance of wine-associated microorganisms. In *Biology of Microorganisms on Grapes, in Must and in Wine*, 2nd ed.; König, H., Unden, G., Fröhlich, J., Eds.; Springer: Heidelberg, Germany, 2017; pp. 549–571.
16. König, H.; Fröhlich, J. Lactic acid bacteria. In *Biology of Microorganisms on Grapes, in Must and in Wine*, 2nd ed.; König, H., Unden, G., Fröhlich, J., Eds.; Springer: Heidelberg, Germany, 2017; pp. 3–41.
17. Guillamón, J.M.; Mas, A. Acetic acid bacteria. In *Biology of Microorganisms on Grapes, in Must and in Wine*, 2nd ed.; König, H., Unden, G., Fröhlich, J., Eds.; Springer: Heidelberg, Germany, 2017; pp. 43–64.
18. Mendes-Ferreira, A.; lí del Olmo, M.; Garcia-Martínez, J.; Pérez-Ortin, J. Functional genomics in wine yeast: DNA arrays and next generation sequencing. In *Biology of Microorganisms on Grapes, in Must and in Wine*, 2nd ed.; König, H., Unden, G., Fröhlich, J., Eds.; Springer: Heidelberg, Germany, 2017; pp. 573–604.
19. Fröhlich, J.; König, H.; Claus, H. Molecular methods for the identification of wine microorganisms and yeast development. In *Biology of Microorganisms on Grapes, in Must and in Wine*, 2nd ed.; König, H., Unden, G., Fröhlich, J., Eds.; Springer: Heidelberg, Germany, 2017; pp. 517–547.
20. Morrison-Whittle, P.; Goddard, M.R. From vineyard to winery: A source map of microbial diversity driving wine fermentation. *Environ. Microbiol.* **2018**, *20*, 75–84. [CrossRef] [PubMed]
21. Borneman, A.R.; Pretorius, I.S.; Chambers, P.J. Comparative genomics: a revolutionary tool for wine yeast strain development. *Curr. Opin. Biotechnol.* **2013**, *24*, 192–199. [CrossRef] [PubMed]
22. Bisson, L.F. Stuck and sluggish fermentations. *Am. J. Enol. Vitic.* **1999**, *50*, 1–13.
23. Bisson, L.F.; Butzke, C.E. Diagnosis and rectification of stuck and sluggish fermentations. *Am. J. Enol. Vitic.* **2000**, *51*, 168–177.
24. Heinisch, J.J.; Rodicio, R. Stress responses in wine yeast. In *Biology of Microorganisms on Grapes, in Must and in Wine*, 2nd ed.; König, H., Unden, G., Fröhlich, J., Eds.; Springer: Heidelberg, Germany, 2017; pp. 377–395.
25. Ribéreau-Gayon, P.; Dubourdieu, D.; Donèche, B.; Lonvaud, A. *Handbook of Enology. The Microbiology of Wine and Vinifications*, 2nd ed.; John Wiley & Sons Ltd.: Chichester, UK, 2006; Volume 1, pp. 79–113.
26. Petruzzi, L.; Capozzi, V.; Berbegal, C.; Corbo, M.R.; Bevilacqua, A.; Spano, G.; Sinigaglia, M. Microbial resources and enological significance: Opportunities and benefits. *Front. Microbiol.* **2017**, *8*, 995. [CrossRef] [PubMed]

27. Jolly, N.P.; Varela, C.; Pretorius, I.S. Not your ordinary yeast: Non-*Saccharomyces* yeasts in wine production uncovered. *FEMS Yeast Res.* **2014**, *14*, 215–237. [CrossRef] [PubMed]

28. Mas, A.; Guillamón, J.M.; Beltran, G. Non-Conventional Yeast in the Wine Industry. *Front. Media* **2016**, *7*, 1494. [CrossRef]

29. Visser, W.; Scheffers, W.A.; Batenburg-van der Vegte, A.H.; van Dijken, J.P. Oxygen requirements of yeasts. *Appl. Environ. Microbiol.* **1990**, *56*, 3785–3792. [PubMed]

30. Hanl, L.; Sommer, P.; Arneborg, N. The effect of decreasing oxygen feed rates on growth and metabolism of *Torulaspora delbrueckii*. *Appl. Microbiol. Biotechnol.* **2005**, *67*, 113–118. [CrossRef] [PubMed]

31. Mauricio, J.C.; Millán, C.; Ortega, J.M. Influence of oxygen on the biosynthesis of cellular fatty acids, sterols and phospholipids during alcoholic fermentation by *Saccharomyces cerevisiae* and *Torulaspora delbrueckii*. *World J. Microbiol. Biotechnol.* **1998**, *14*, 405–410. [CrossRef]

32. Brandam, C.; Lai, Q.P.; Julien-Ortiz, A.; Taillandier, P. Influence of oxygen on alcoholic fermentation by a wine strain of *Torulaspory delbrueckii*: Kinetics and carbon mass balance. *Biosci. Biotechnol. Biochem.* **2013**, *77*, 1848–1853. [CrossRef] [PubMed]

33. Dimopoulou, M.; Lonvaud-Funel, A.; Dols-Lafargue, M. Polysaccharide production by grapes must and wine microorganisms. In *Biology of Microorganisms on Grapes, in Must and in Wine*, 2nd ed.; König, H., Unden, G., Fröhlich, J., Eds.; Springer: Heidelberg, Germany, 2017; pp. 293–314.

34. Sanz, L.M.; Martínez-Castro, I. Carbohydrates. In *Wine Chemistry and Biochemistry*; Moreno-Arribas, M.V., Polo, M.C., Eds.; Springer: Heidelberg, Germany, 2010; pp. 231–248.

35. Santos-Buelga, C.; de Freitas, V. Influence of phenolics on wine organoleptic properties. In *Wine Chemistry and Biochemistry*; Moreno-Arribas, M.V., Polo, M.C., Eds.; Springer: Heidelberg, Germany, 2010; pp. 529–570.

36. Sabel, A.; Claus, H.; Martens, S.; König, H. *Wickerhamomyces anomalus* AS1: A new strain with potential to improve wine aroma. *Ann. Microbiol.* **2014**, *64*, 483–491. [CrossRef]

37. Schwentke, J.; Sabel, A.; Petri, A.; König, H.; Claus, H. The yeast *Wickerhamomyces anomalus* AS1 secretes a multifunctional exo-ß-1,3-glucanase with implications for winemaking. *Yeast* **2014**, *31*, 349–359. [CrossRef] [PubMed]

38. Dittrich, H.H.; Großmann, M. *Mikrobiologie des Weines. 4. aktualisierte Auflage*; Ulmer Verlag: Stuttgart, Germany, 2010.

39. Christ, E.; Kowalczyk, M.; Zuchowska, M.; Claus, H.; Löwenstein, R.; Szopinska-Morawska, A.; Renaut, J.; König, H. An exemplary model study for overcoming stuck fermentation during spontaneous fermentation with the aid of a *Saccharomyces* triple hybrid. *J. Agric. Sci.* **2015**, *7*, 18–34. [CrossRef]

40. Szopinska, A.; Christ, E.; Planchon, S.; König, H.; Evers, D.; Renaut, J. Stuck at work? Quantitative proteomics of environmental wine yeast strains reveals the natural mechanism of overcoming stuck fermentation. *Proteomics* **2016**, *16*, 593–608. [CrossRef] [PubMed]

41. Blättel, V.; Petri, A.; Rabenstein, A.; Kuever, J.; König, H. Differentiation of species of the genus *Saccharomyces* using biomolecular fingerprinting methods. *Appl. Microbiol. Biotechnol.* **2013**, *97*, 4597–4606. [CrossRef] [PubMed]

42. Jiranek, A.; Langridge, P.; Henschke, P.A. Amino acid and ammonium utilization by *Saccharomyces cerevisiae* wine yeasts from a chemically defined medium. *Am. J. Enol. Vitic.* **1995**, *46*, 75–83.

43. Zuchowska, M.; König, H.; Claus, H. Allelic variants of hexose transporter Hxt3p und hexokinase Hxk1p/Hxk2p in strains of *Saccharomyces cerevisiae* und interspecies hybrids. *Yeast* **2015**, *32*, 657–669. [CrossRef] [PubMed]

44. González, S.S.; Barrio, E.; Gafner, J.; Querol, A. Natural hybrids from *Saccharomyces* cerevisiae, *Saccharomyces bayanus* and *Saccharomyces kudriavzevii* in wine fermentations. *FEMS Yeast Res.* **2006**, *6*, 1221–1234. [CrossRef] [PubMed]

45. Bradbury, J.; Richards, K.; Niederer, H.; Lee, S.; Rod Dunbar, P.; Gardner, R. A homozygous diploidsubset of commercial wine yeast strains. *Antonie van Leeuwenhoek* **2006**, *89*, 27–37. [CrossRef] [PubMed]

46. González, S.S.; Barrio, E.; Querol, A. Molecular characterization of new natural hybrids between *Saccharomyces cerevisiae* and *Saccharomyces kudriavzevii* from brewing. *Appl. Environ. Microbiol.* **2008**, *74*, 2314–2320. [CrossRef] [PubMed]

47. Lopes, C.A.; Barrio, E.; Querol, A. Natural hybrids of *S. cerevisiae* × *S. kudriavzevii* share alleles with European wild populations of *Saccharomyces kudriavzevii*. *FEMS Yeast Res.* **2010**, *10*, 412–421. [CrossRef] [PubMed]

48. Naumov, G.I.; James, S.A.; Naumova, E.S.; Louis, E.J.; Roberts, I.N. Three new species in the *Saccharomyces sensu stricto* complex: *Saccharomyces cariocanus, Saccharomyces kudriavzevii* and *Saccharomyces mikatae*. *Int. J. System. Evol. Microbiol.* **2000**, *50*, 1931–1942. [CrossRef] [PubMed]

49. Sampaio, J.P.; Gonçalves, P. Natural populations of *Saccharomyces kudriavzevii* in Portugal are associated with oak bark and sympatric with *S. cerevisiae* and *S. paradoxus*. *Appl. Environ. Microbiol.* **2008**, *74*, 2144–2152. [CrossRef] [PubMed]

50. Hirschhäuser, S.; Fröhlich, J.; Gneipel, A.; Schönig, I.; König, H. Fast protocols for the 5S rDNA and ITS-2 based identification of *Oenococcus oeni*. *FEMS Lett.* **2005**, *244*, 165–171. [CrossRef] [PubMed]

51. Sebastian, P.; Herr, P.; Fischer, U.; König, H. Molecular identification of lactic acid bacteria occurring in must and wine. *South Afr. J. Enol. Vitic.* **2011**, *32*, 300–309. [CrossRef]

52. Röder, C.; König, H.; Fröhlich, J. Species-specific identification of *Dekkera/Brettanomyces* yeasts by fluorescently labelled rDNA probes targeting the 26S rRNA. *FEMS Yeast Res.* **2007**, *7*, 1013–1026. [CrossRef] [PubMed]

53. Petri, A.; Pfannebecker, J.; Fröhlich, J.; König, H. Fast identification of wine related lactic acid bacteria by multiplex PCR. *Food Microbiol.* **2013**, *33*, 48–54. [CrossRef] [PubMed]

54. Chambers, P.J.; Pretorius, I.S. Fermenting knowledge: The history of winemaking, science and yeast research. *EMBO Rep.* **2010**, *11*, 914–920. [CrossRef] [PubMed]

55. Pretorius, I.S. Tailoring wine yeast for the new millennium: Novel approaches to the ancient art of wine making. *Yeast* **2000**, *16*, 675–729. [CrossRef]

56. Borneman, A.R.; Desany, B.A.; Riches, D.; Affourtit, J.P.; Forgan, A.H.; Pretorius, I.S.; Egholm, M.; Chambers, P.J. The genome sequence of the wine yeast VIN7 reveals an allotriploid hybrid genome with *Saccharomyces cerevisiae* and *Saccharomyces kudriavzevii* origins. *FEMS Yeast Res.* **2012**, *12*, 88–96. [CrossRef] [PubMed]

fermentation

MDPI

Article

Characterization of *Saccharomyces bayanus* CN1 for Fermenting Partially Dehydrated Grapes Grown in Cool Climate Winemaking Regions

Jennifer Kelly [1], Fei Yang [2], Lisa Dowling [2], Canan Nurgel [3], Ailin Beh [3], Fred Di Profio [2], Gary Pickering [2,3] and Debra L. Inglis [1,2,3,]*

[1] Centre for Biotechnology, Brock University, St. Catharines, ON L2S 3A1, Canada; jk13wk@brocku.ca
[2] Cool Climate Oenology and Viticulture Institute, Brock University, St. Catharines, ON L2S 3A1, Canada; fyang2@brocku.ca (F.Y.); ldowling@brocku.ca (L.D.); fdiprofio@yahoo.ca (F.D.P.); gpickering@brocku.ca (G.P.)
[3] Department of Biological Sciences, Brock University, St. Catharines, ON L2S 3A1, Canada; Canan.Nurgel@loblaw.ca (C.N.); ailin.beh@gmail.com (A.B.)
*   Correspondence: dinglis@brocku.ca; Tel.: +1-905-688-5550 (ext. 3828)

Received: 15 August 2018; Accepted: 11 September 2018; Published: 13 September 2018

Abstract: This project aims to characterize and define an autochthonous yeast, *Saccharomyces bayanus* CN1, for wine production from partially dehydrated grapes. The yeast was identified via PCR and Basic Local Alignment Search Tool (BLAST) analysis as *Saccharomyces bayanus*, and then subsequently used in fermentations using partially dehydrated or control grapes. Wine grapes were dried to 28.0° Brix from the control grapes at a regular harvest of 23.0° Brix. Both the partially dehydrated and control grapes were then vinified with each of two yeast strains, *S. bayanus* CN1 and *S. cerevisiae* EC1118, which is a common yeast used for making wine from partially dehydrated grapes. Chemical analysis gas chromatography-flame ionization detector (GC-FID) and enzymatic) of wines at each starting sugar level showed that CN1 produced comparable ethanol levels to EC1118, while producing higher levels of glycerol, but lower levels of oxidative compounds (acetic acid, ethyl acetate, and acetaldehyde) compared to EC1118. Yeast choice impacted the wine hue; the degree of red pigment coloration and total red pigment concentration differed between yeasts. A sensory triangle test ($n = 40$) showed that wines made from different starting sugar concentrations and yeast strains both differed significantly. This newly identified *S. bayanus* strain appears to be well-suited for this style of wine production from partially dehydrated grapes by reducing the oxidative compounds in the wine, with potential commercial application for cool climate wine regions.

Keywords: winemaking; partially dehydrated grapes; appassimento; yeast; *Saccharomyces bayanus*; sensory; Ontario; climate change adaptation

1. Introduction

In an increasingly competitive international marketplace, important strategic considerations include a focus on the reliable production of high-value wines, and on styles that help differentiate and brand a wine region. This creates particular opportunities for the emerging wine regions of the New World, to adapt the traditions of the Old World while developing technological advancements in viticulture and oenology to assist in the expression of regionality [1]. In the recent past, winemakers in Ontario, Canada have highlighted their unique regional identity with products such as sparkling Icewine (e.g., Inniskillin Wines). Moving beyond that, there is room for additional signature products that can help define this region. Developing such wine styles and their corresponding production technologies can support the sustainability of established appellations, as well as the development of nascent grape-growing regions.

The Ontario industry is economically important [2], and its success is intrinsically linked to its unique climate, which allows the growth of a range of premium *vinifera* grape varieties [3]. However, it can be challenging to achieve optimal grape ripeness in the shorter growing season that is associated with Ontario's cool climate [4]. Further, weather volatility is an additional threat to grape-growing in this region, with the most salient risks associated with temperature extremes, rainfall variability, and winter and frost damage [5]. Therefore, it is prudent to adopt innovative strategies in order to mitigate the risks associated with a changing climate and stabilize quality from vintage to vintage.

Postharvest grape-drying (appassimento) followed by vinification is a technique that is traditionally employed in Northern Italy for Amarone wine production [6]. This method consists of ripening grapes off-vine to produce withered or partially dehydrated fruit. The drying process increases the concentration of total soluble solids, phenolic compounds, and odorants in the grapes [7,8]. The wines produced from these grapes have a higher concentration of ethanol, volatile aroma compounds, and anthocyanins [9,10]. In Ontario, Canada, wines made from partially dehydrated grapes are regulated by the Vintners Quality Alliance (VQA) under the term Vin de Curé [11].

Despite these benefits, wines made from partially dehydrated grapes can have increased levels of undesirable oxidation compounds in the wine, most notably acetic acid, ethyl acetate, and acetaldehyde [10,12,13]. At elevated concentrations, these compounds can negatively affect the organoleptic quality of the wine [14], and in the case of acetic acid, exceed legal limits enforced by the VQA [11]. The development of these compounds is directly related to the high starting sugar concentration in the must that creates an environment of high osmotic stress for yeast.

Glycerol, the major compatible solute in *S. cerevisiae*, accumulates intracellularly as a survival response to hyperosmotic stress [15]. The accumulation of glycerol maintains cell volume and turgor pressure while limiting the efflux of intracellular water [15,16]. Glycerol formation is accompanied by an increase in NAD^+ production [17]. Under these conditions, the shift in redox balance ($NADH:NAD^+$ ratio) caused by the increased formation of glycerol is corrected via acetic acid production, which reduces NAD^+ to NADH [17–20]. Monitoring the development of glycerol and acetic acid during fermentation can therefore provide insights into the yeast's management of redox balance and hyperosmotic stress.

It has been suggested that autochthonous starter cultures have benefits for regional wines, including sparkling wines, in that they may be well-adapted to specific environmental conditions, and prospectively enhance the desired flavor and aroma profiles, which can impact the quality of regional wines [21–26]. We previously conducted a spontaneous fermentation of local Riesling Icewine must from Ontario, Canada and identified that *Candida dattilla* along with *Kloeckera apiculata* and *Cryptococcus laurentii* dominated the fermentation, and were still present at the end (day 30), whereas *S. cerevisiae* was not found [27]. In a later study, this *Candida dattilla* strain, which was initially identified using API Biomedical kits, was further identified as a *Saccharomyces* species by DNA sequencing of the 5.8S-ITS region. It was likely *S. bayanus* or *S. pastorianus*, but the identification could not be finalized past the genus (unpublished). Since *S. bayanus* is reported as producing lower acetic acid levels during wine fermentation [28], the strain isolated from Icewine grapes in Ontario was further tested on its own in the osmotically stressful Icewine fermentation condition. A pure starter culture of this yeast was built up and inoculated into filter-sterilized 41.6° Brix Riesling Icewine juice, where it produced 7.7% *v/v* ethanol compared to 10.8% *v/v* from the control *S. cerevisiae* K1-V1116. However, the isolated yeast produced 1.3-fold lower acetic acid/sugar consumed compared to K1-V1116 [29]. Although this yeast did attain the minimum alcohol required for Icewine of 7%, commercial Icewines in Canada have been found to range between 8.4–12.6% *v/v* ethanol and for Riesling Icewines, between 9.1–12.2% *v/v* [30]. The combined value of autochthonous yeast for the expression of regionality and the positive preliminary results in Icewine led us to characterize this yeast strain during the fermentation of must from partially dehydrated grapes, which provides a less stressful sugar environment than Icewine juice, but still has potentially problematic oxidative quality concerns from this wine style [10,12,13].

In this study, a local yeast isolated from the skin of Riesling Icewine grapes [27] is tested in the fermentation of partially dehydrated grapes. Grapes were dried to 28.0°Brix and vinified with one of two yeast strains, *S. cerevisiae* EC1118, the commonly used yeast for this wine style, and the yeast of interest, CN1. Grapes picked at 23.0°Brix (a sugar level typical for red table wine production) were also fermented with the two yeast strains as a control.

The main objectives of our study are to (i) identify this locally-isolated yeast, (ii) determine its fitness for making wine from partially dehydrated grapes, and (iii) more fully understand the impact of high sugar fermentation on red wine composition, color, and sensory quality. The results from this study should assist in optimizing winemaking from partially dehydrated grapes in cool climate wine areas such as Ontario, Canada, as well as inform international wine regions that are seeking regional differentiation or further innovation of their wine styles.

2. Materials and Methods

2.1. Yeast Strains

Two yeast strains were selected to carry out the fermentations. The commercial *S. cerevisiae* strain EC1118, was purchased from Lallemand (Montreal, QC, Canada). The local strain was isolated from Riesling Icewine grapes from the Niagara Region in Ontario, at the Cool Climate Oenology and Viticulture Institute (CCOVI). Four genomic areas were analyzed to identify this yeast: the internal transcribed spacer regions (ITS1 and ITS2), including the *5.8S* gene of the ribosomal DNA (GenBank accession number: MH317189); the D1/D2 domain of a large subunit of the *26S rRNA* gene region (GenBank accession number: MH318011); the mitochondrial *β-tubulin* gene (GenBank accession number: MH339593); and the mitochondrial cytochrome oxidase II gene (*COXII*) (GenBank accession number: MH339594). The ITS1-*5.8S rRNA*-ITS2 gene region was amplified via PCR with the universal primers ITS1 (5′-TCCGTAGGTGAACCTGCGG) and ITS4 (5′-TCCTCCGCTTATTGATATGC). The D1/D2 domain was amplified with the primers NL-1 (5′-GCATATCAATAAGCGGAGGAAAAG) and NL-4 (5′-GGTCCGJGTTTCAAGACGG). The *β-tubulin* gene was amplified with the primer pair βtub3 (5′-TGGGCYAAGGGTYAYTAYAC) and βtub4r (5′-GCCTCAGTRAAYTCCATYTCRTCCAT), and the *COXII* gene was amplified with the primers COII5 (5′-GGTATTTTAGAATTACATGA) and COII3 (5′-ATTTATTGTTCRTTTAATCA). DNA sequencing analysis (Robarts Research Institute, London, ON, Canada) was performed on all four amplified genes, and the results were compared with all of the available sequence databases of DNA using the Basic Local Alignment Search Tool (BLAST).

2.2. Grape Harvest, Desiccation and Processing

Vitis vinifera Cabernet franc grapes were hand-harvested at Mazza Vineyards in Niagara-on-the-Lake, Ontario, Canada, at approximately 23.0°Brix. First, 209 kg of grapes were picked and placed in perforated drying containers in a single layer. Grapes were divided into two parcels and delivered to two locations. One of the parcels was delivered to CCOVI (Brock University, St Catharines, ON, Canada) and processed on the following day after temperature stabilization overnight at room temperature. The other parcel was delivered to Cave Spring Cellars Barn (4424 Cave Spring Road, Beamsville, ON, Canada), which is dedicated to drying grapes for producing commercial Vin de Curé wines [11]. The drying containers were stacked 14 layers high, with adequate air space between each container to receive natural ventilation in the barn. Fifteen randomly selected clusters were collected weekly. The samples were hand-crushed in a plastic bag and strained through a metal strainer to collect must. Must samples were analyzed for soluble solids, pH, and titratable acidity. Once the target sugar concentration was reached (28.0°Brix), the partially dehydrated grapes were delivered to CCOVI for processing after temperature stabilization overnight. Grapes were crushed and destemmed (model Gamma 50, Mori-TEM; Florence, Italy) into 30-L steel fermentation vessels with tight-fitting lids. Must was blanketed with CO_2, lids were secured, and vessels were stored at 22 °C prior to yeast inoculation. Must volume was estimated by multiplying weight by 0.75 for control must, and 0.60 for partially dehydrated grape must to account for desiccation effects. Then, 500 mg L^{-1} of diammonium phosphate (DAP; Laffort, Bordeaux, France) was added to the must and mixed by punch down. A further 250 mg L^{-1} of DAP was added on the third day of fermentation to reduce yeast stress.

2.3. Winemaking

Four sets of triplicate fermentations were carried out: (i) 23.0°Brix must fermented with *S. cerevisiae* EC1118, (ii) 23.0°Brix must fermented with *S. bayanus* CN1, (iii) 28.0°Brix must fermented with EC1118, and (iv) 28.0°Brix must fermented with CN1. Fermentations were conducted using the same microvinification protocols. *S. cerevisiae* EC1118 was rehydrated according to manufacturer's directions and plated out on yeast extract peptone dextrose plates (YPD, 1% yeast extract, 2% peptone, 2% dextrose, 2% agar). CN1 yeast was prepared from a frozen glycerol stock, and also plated out on YPD plates. Both yeasts were grown to appropriate colony size prior to preparing a starter culture in sterile-filtered grape juice. The starter cultures were built up in sterile-filtered Cabernet franc must, and then followed a step-wise acclimatization procedure as outlined in Kontkanen et al. [20]. The yeast strains were inoculated from YPD plates into 750 mL of 10°Brix sterile-filtered must with the addition of 2 g L^{-1} DAP and grown aerobically at 25 °C with shaking at 0.605× *g* until cell concentration reached 2×10^8 cells mL^{-1}, as determined by haemocytometry. Then, 750 mL of sterile-filtered 23.0°Brix control must was added to each build-up culture and held for 1 h at 25 °C with swirling every half hour. The 1.5 L of control cultures for both EC1118 and CN1 were added to 28.5 L of 23°Brix control must to reach an inoculum of 5.0×10^6 cells mL^{-1} in 30-L stainless steel fermentation vessels. The 28.0°Brix treatment required one more acclimatization step for both yeast, and 750 mL of sterile-filtered 28.0°Brix dehydrated grape must was added to each starter culture and held for 2 h at 25 °C with swirling every half hour, after which the 2.25-L culture was inoculated into 27.75 L of 28.0°Brix dehydrated grape must to reach an inoculum of 5.0×10^6 cells mL^{-1} in the 30-L fermentations.

After inoculation, the fermentations were gently mixed by punch down and moved to a temperature-controlled chamber at 22 °C. Fermentations were monitored once daily by recording soluble solids (hydrometer, °Brix) and temperature (thermometer, °C). The caps were punched down twice daily with 20 plunges per vessel using a separate punch-down tool for each yeast trial; this number was gradually reduced to four plunges near the end of the fermentation. As the cap started to fall, fermentations were blanketed with CO_2 to protect them from oxidation. Fermentations were considered complete once the yeast stopped consuming sugar (<5 g L^{-1}) and/or the sugar concentration stayed the same for three consecutive days, as confirmed by a wine scan analysis conducted by WineScan™ FT120 (FOSS, Hillerød, Denmark). Once complete, fermentation replicates were pressed separately with a small bladder press (Enotecnica Pillan, Vicenza, Italy) at 1 bar for 2 min into glass carboys. Then, 50 mg L^{-1} of sulfur dioxide (as potassium metabisulfite) was added to each treatment, which were left to settle at room temperature. Wines were then racked and moved to a −2 °C chamber for cold stabilization. Wines were subsequently filtered through 0.45-µm filter pads, bottled in 750-mL glass wine bottles, with a manual bottler (Criveller Group; Niagara Falls, ON, Canada), closed with natural cork with an automated corker (model ETSILON-R, Bertolaso; San Vito, Italy), and stored in the CCOVI wine cellar (17.5 °C, 74.5% RH).

2.4. Grape, Must, and Fermentation Analysis

Fermentation temperature was monitored with a thermometer (°C). Soluble solids were determined using an Abbe bench top refractometer (model 10450, American Optical; Buffalo, NY, USA) for grape and must samples, and using a degree Brix hydrometer for fermentation time course samples. pH was determined using a pH meter (SympHony, VWR, SB70P, Mississauga, ON, Canada), and titratable acidity was determined by titration with 0.1 mol L^{-1} of NaOH to an endpoint of pH 8.2 [31]. Glucose, fructose, glycerol, acetaldehyde, ethanol in must, amino acid nitrogen, ammonia nitrogen, acetic acid, lactic acid, and malic acid were determined with Megazyme Kits

(K-FRUGL, K-GCROL, K-ACHD, K-ETOH, K-PANOPA, K-AMIAR, K-ACET, K-LATE, K-LMALL; Megazyme International Ireland, Limited, Bray Company, Wicklow, Ireland). Ethyl acetate and ethanol in wine were determined by gas chromatography (GC) using a Hewlett-Packard 6890 series gas chromatograph (Agilent Technologies Incorporated, Santa Clara, CA, USA) equipped with a flame ionization detector (FID), split/split-less injector, and Chemstation software (version E.02.00.493). Separations were carried out with a DB®-WAX (30 m, 0.25 mm, 0.25 μm) GC column (122-7032 model; Agilent Technologies, Santa Clara, CA, USA) with helium as the carrier gas at a flow rate of 1.5 mL min^{-1}.

2.5. Color Evaluation

Measures of color density, hue, degree of red pigment coloration, and total red pigments were conducted based on the methods of Iland et al. [32] by UV-Vis spectrophotometer (Cary 60, Agilent Technologies, Santa Clara, CA, USA).

2.6. Sensory Evaluation

A preliminary bench tasting ($n = 4$) of the wines established that the winemaking replicates within each treatment were similar enough to blend into representative treatments for difference testing. Therefore, four treatments were presented to the panelists (EC1118, 23.0°Brix; CN1, 23.0°Brix; EC1118, 28.0°Brix; CN1, 28.0°Brix). A balanced and randomized triangle test design composed of six sets of triads was used to compare all of the treatments to each other. Each participant ($n = 40$) tasted a total of 18 samples over the course of two sessions. The first session consisted of three sets of three wines, separated by forced three-minute breaks between each set to minimize fatigue and carry-over effects. Consumption of water and unsalted crackers was encouraged. The samples were coded with a three-digit randomly assigned code, and the participants were asked to evaluate them in the order presented. Participants were instructed to assess aroma by sniffing and flavor by tasting and expectorating the samples, and determine differences based on these observations. Their answers were recorded using the Compusense Five™ computer program (Compusense Inc., Guelph, ON, Canada). The same format was used for the second session, which was completed after a one-hour break. The evaluations took place in individual booths in the sensory evaluation lab at CCOVI, which was equipped with red lighting to mask possible color differences. Data was analyzed by comparing the number of correct responses to a critical value table for triangle tests [33].

2.7. Statistical Analysis

Analysis of variance (ANOVA) with mean separation by Fisher's Protected Least Significant Difference (LSD) test ($p < 0.05$) was conducted on chemical and color parameters using the XLSTAT statistical software package (Addinsoft, Version 7.1; New York, NY, USA).

2.8. Statement of Ethics

All of the subjects gave their informed consent for inclusion before they participated in the study. The protocol for the study was approved by Brock University's Research Ethics Board (file number 14-021-INGLIS).

3. Results

3.1. Yeast Strain Identification

The sequencing results of the ITS1-*5.8S rRNA*-ITS2 gene region and the D1/D2 domain gene region were only able to identify the isolate at the genus level as a *Saccharomyces* strain. Therefore, the mitochondrial genes *β-tubulin* [34] and *COXII* [35] were selected as biomarkers for further identification. The amplified sequences of *β-tubulin* showed a 99% similarity in sequence identity with a query coverage of 100% to three *S. bayanus* strains (Table 1). The results from the *COXII* mitochondrial gene reported an identical level of similarity to CBS 380[T] and CBS 395[T] (Table 1), which are widely accepted type strains (taxonomic standards) of *S. bayanus* and *S. uvarum*, respectively [36,37]. Based on the Genbank sequence comparisons, we have identified this yeast as *S. bayanus*. Recent research reports the nearly identical similarity of the complete mitochondrial genome between these two potential species [38], further raising the question of whether *S. bayanus* and *S. uvarum* should be classified into two separate species (*S. bayanus*, *S. uvarum*) or two varieties under the species *S. bayanus* (*S. bayanus* var. *bayanus*, *S. bayanus* var. *uvarum*) [36].

Table 1. Homology of CN1 mitochondrial genes with GenBank sequences.

Gene Region	NCBI Database Strain for Sequence Comparison	GenBank Accession Number	Base Pairs *	Alignment Results		
				Max Score	Query Coverage	Sequence Identity
β-tubulin	*S. bayanus* Strain BCRC 21818	FJ238317.1	849/852	1555	100%	99%
	S. bayanus Strain BCRC 21964	FJ238319.1	848/852	1550	100%	99%
	S. bayanus Strain BCRC 21816	FJ238316.1	847/852	1546	100%	99%
	S. eubayanus Strain N/A	XM 018364800.1	815/852	1367	100%	96%
	S. pastorianus Strain BCRC 21420	FJ238324.1	813/852	1356	100%	95%
COXII	*S. bayanus* Strain CBS380[T]	KX657743.1	632/635	1157	99%	99%
	S. uvarum Strain CBS395[T]	KX657742.1	632/635	1157	99%	99%
	S. bayanus Strain CBS380	AP014933.1	632/635	1157	99%	99%
	S. bayanus x *S. uvarum* Strain CECT1991	JN676774.1	585/585	1081	91%	100%
	S. eubayanus Strain CRUB1975	KF530344.1	608/620	1079	97%	98%

* Number of base pairs in common between amplified sample and NCBI database sequence/total base pairs aligned.

3.2. Fermentation Kinetics and Metabolites

The must parameters for all treatments are listed in Table 2. The *S. bayanus* CN1 yeast consumed sugars at a higher rate than the control yeast EC1118 at the beginning of both fermentation treatments, but left 15.8 g L^{-1} unfermented sugar (mainly fructose) in the 28°Brix treatment wine (Figure 1, Table 3). Despite CN1 leaving residual sugar in the high brix ferments, CN1 produced a comparable level of ethanol to EC1118, and significantly less oxidative compounds (acetaldehyde, acetic acid, ethyl acetate) for both the control and wines made from the dehydrated grapes (Table 3). Regardless of the winemaking treatment, wines fermented with CN1 contained higher levels of glycerol, titratable acidity, and malic acid in comparison to wines fermented with EC1118, but lower lactic acid in the 23°Brix fermentation (Table 3).

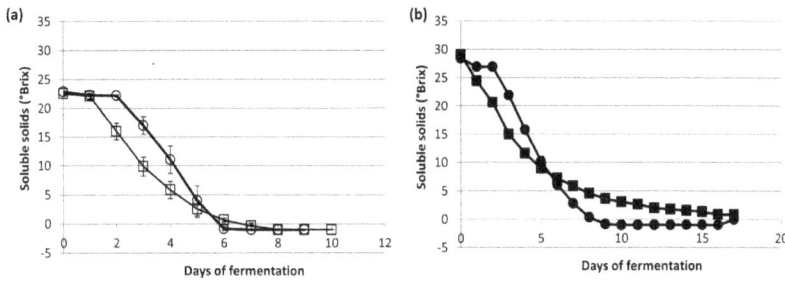

Figure 1. Soluble solid levels during fermentation. (**a**) The 23°Brix control must was inoculated with EC1118 (○) and CN1 (□); (**b**) the 28°Brix partially dehydrated grape must was inoculated with EC1118 (●) and CN1 (■). Data represents the mean value ± standard deviation of duplicate measurements per sample (three winemaking replicates per treatment).

Table 2. Chemical composition of Cabernet franc control must (23°Brix) and must from partially dehydrated grapes (28°Brix). Data represents the mean value ± standard deviation of duplicate measurements per sample (three winemaking replicates per treatment). Lowercase letters within the same parameter indicate differences between treatments (Fisher's Protected Least Significant Difference (LSD)$_{0.05}$).

Parameter	23°Brix EC1118	23°Brix CN1	28°Brix EC1118	28°Brix CN1
Reducing sugar (g L^{-1})	218 ± 8 [b]	198 ± 12 [a]	300 ± 3 [c]	301 ± 3 [c]
Glucose (g L^{-1})	108 ± 4 [b]	98 ± 6 [a]	145 ± 2 [c]	145 ± 1 [c]
Fructose (g L^{-1})	111 ± 5 [b]	100 ± 6 [a]	155 ± 2 [c]	156 ± 2 [c]
pH	3.39 ± 0.05 [a]	3.35 ± 0.01 [a]	3.34 ± 0.03 [a]	3.33 ± 0.03 [a]
Titratable acidity (g L^{-1} tartaric acid)	5.8 ± 0.2 [b]	6.1 ± 0.1 [c]	4.8 ± 0.0 [a]	4.9 ± 0.0 [a]
Ammonia nitrogen (mg N L^{-1})	17 ± 9 [b]	12 ± 2 [a,b]	8 ± 1 [a]	8 ± 2 [a,b]
Primary amino nitrogen (mg N L^{-1})	62 ± 13 [b]	47 ± 2 [a]	61 ± 3 [b]	63 ± 5 [b]
Ethanol (% v/v)	0.009 ± 0.004 [a]	0.005 ± 0.001 [a]	0.030 ± 0.006 [b]	0.031 ± 0.006 [b]
Glycerol (g L^{-1})	0.0 ± 0.0 [a]	0.0 ± 0.0 [a]	0.3 ± 0.1 [b]	0.3 ± 0.0 [b]
Malic acid (g L^{-1})	2.2 ± 0.3 [a]	2.1 ± 0.1 [a]	2.1 ± 0.1 [a]	2.0 ± 0.1 [a]
Lactic acid (g L^{-1})	0.04 ± 0.00 [a]	0.04 ± 0.11 [a]	0.05 ± 0.00 [b]	0.06 ± 0.00 [b]
Acetaldehyde (mg L^{-1})	<18 [a]	<18 [a]	<18 [a]	<18 [a]
Acetic acid (g L^{-1})	0.01 ± 0.00 [a]	0.00 ± 0.00 [a]	0.01 ± 0.00 [b]	0.01 ± 0.00 [b]
Ethyl acetate (mg L^{-1})	n/d †	n/d †	n/d †	n/d †

† n/d indicates the measurement is not detectable.

Table 3. Chemical composition of Cabernet franc control wine (23°Brix) and wine made from partially dehydrated grapes (28°Brix). Data represents the mean value ± standard deviation of duplicate measurements per sample (three winemaking replicates per treatment). Lowercase letters within the same parameter indicate differences between treatments (Fisher's Protected LSD$_{0.05}$).

Parameter	23°Brix EC1118	23°Brix CN1	28°Brix EC1118	28°Brix CN1
Reducing sugar (g L^{-1})	<0.07 [a]	0.2 ± 0.0 [a]	<0.07 [a]	15.8 ± 6.7 [b]
Glucose (g L^{-1})	<0.07 [a]	<0.07 [a]	<0.07 [a]	1.1 ± 0.7 [b]
Fructose (g L^{-1})	<0.07 [a]	0.1 ± 0.0 [a]	<0.07 [a]	14.7 ± 6.0 [b]
pH	3.78 ± 0.09 [b]	3.54 ± 0.04 [a]	3.74 ± 0.00 [b]	3.59 ± 0.05 [a]
Titratable acidity (g L^{-1} tartaric acid)	6.4 ± 0.3 [a]	9.4 ± 0.3 [c]	6.8 ± 0.2 [a]	8.1 ± 0.3 [b]
Ammonia nitrogen (mg N L^{-1})	<6 [a]	<6 [a]	<6 [a]	<6 [a]
Primary amino nitrogen (mg N L^{-1})	28 ± 3 [a]	24 ± 3 [a]	40 ± 2 [b]	36 ± 4 [b]
Ethanol (% v/v)	13.0 ± 0.3 [a]	12.6 ± 0.4 [a]	15.3 ± 0.7 [b]	14.7 ± 0.2 [b]

Table 3. *Cont.*

Parameter	23°Brix EC1118	23°Brix CN1	28°Brix EC1118	28°Brix CN1
Glycerol (g L^{-1})	8.5 ± 0.4 [a]	11.1 ± 0.6 [b]	11.2 ± 0.1 [b]	13.6 ± 0.2 [c]
Malic acid (g L^{-1})	1.6 ± 0.4 [a]	4.2 ± 0.2 [c]	1.9 ± 0.1 [a]	2.5 ± 0.1 [b]
Lactic acid (g L^{-1})	0.45 ± 0.42 [b]	0.04 ± 0.01 [a]	<0.03 [a]	<0.03 [a]
Acetaldehyde (mg L^{-1})	56 ± 7 [b]	38 ± 5 [a]	88 ± 7 [d]	70 ± 9 [c]
Acetic acid (g L^{-1}) Ethyl acetate (mg L^{-1})	0.30 ± 0.02 [c] 36 ± 3 [b]	0.06 ± 0.01 [a] 21 ± 3 [a]	0.36 ± 0.02 [d] 37 ± 13 [b]	0.20 ± 0.02 [b] 33 ± 2 [a]

3.3. Color and Sensory Evaluation

There were no significant differences between the wines in color density, which describes the intensity of wine color (Figure 2a). The hue, which is a measure of the shade of wine color, was lower in the 23°Brix CN1 wine (Figure 2b). The total red pigments in CN1 wines were lower than that in EC1118 wines for both winemaking treatments (Figure 2c). However, the degree of red pigment coloration was higher in CN1 wines than in EC1118 wines, suggesting a higher percentage of red-colored pigments in wines fermented by CN1 despite the lower concentrations of total red pigments (Figure 2c,d). Sensory evaluation also indicated perceptible differences between all of the wines with different yeast and starting sugar treatment (Table 4).

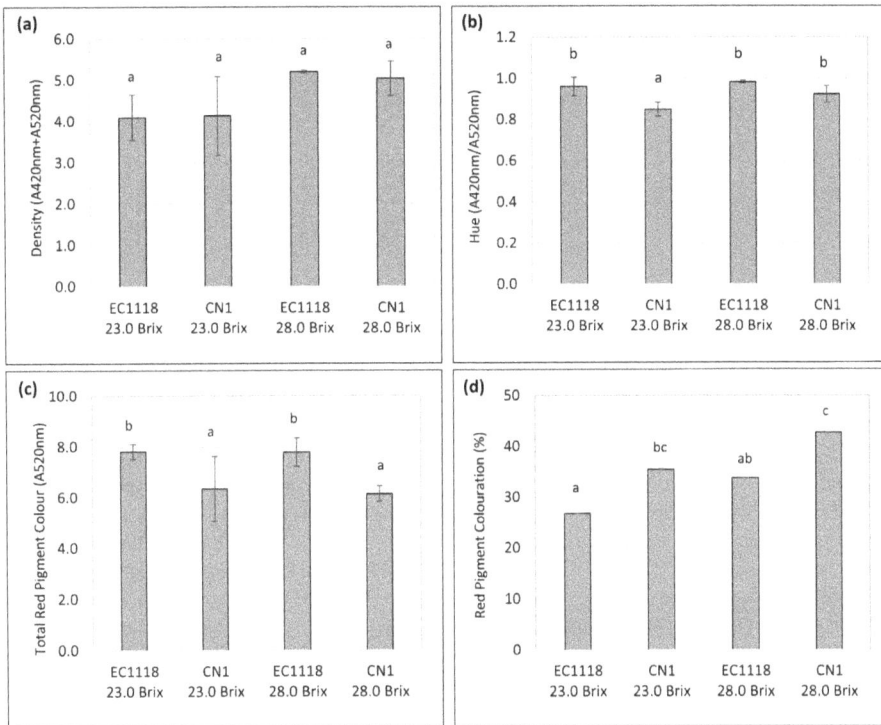

Figure 2. (a) Wine color density (b) wine hue, (c) total red pigment color (anthocyanins, oligomers, and polymers) and (d) degree of red pigment coloration (%) of control wines (23°Brix) and wines made from partially dehydrated grapes (28°Brix) vinified with either EC1118 or CN1. Data represents the mean value ± standard deviation of duplicate measurements per sample (three winemaking replicates per treatment). Lowercase letters indicate differences between treatments (Fisher's Protected LSD$_{0.05}$).

Table 4. Triangle test results to determine sensory differences between wines (n = 40). Significance was assessed by comparing the proportion of correct responses to critical values [33].

Paired Treatments	Correct	Incorrect	Total	Significance
23°Brix EC1118 vs. 23°Brix CN1	25	15	40	p = 0.001
28°Brix EC1118 vs. 28°Brix CN1	34	6	40	p = 0.001
23°Brix EC1118 vs. 28°Brix EC1118	25	15	40	p = 0.001
23°Brix EC1118 vs. 28°Brix CN1	32	8	40	p = 0.001
23°Brix CN1 vs. 28°Brix EC1118	26	14	40	p = 0.001
23°Brix CN1 vs. 28°Brix CN1	37	3	40	p = 0.001

4. Discussion

The main aim of this study is to investigate a low acetic acid-producing yeast, the newly identified yeast *S. bayanus* CN1, within the context of wine production from partially dehydrated grapes, which is a process that involves a high sugar fermentation and is often associated with undesirable oxidation compounds. The results presented in this study are based on chemical and preliminary sensorial analysis that demonstrate lower oxidation compounds produced by CN1 and perceptive differences from EC1118, which is the commonly used yeast for this winemaking style.

In an analysis of Amarone vinified with *S. cerevisiae* EC1118, the authors report concentrations of 0.56 ± 0.02 g L^{-1} acetic acid, 57.20 ± 2.12 mg L^{-1} ethyl acetate, 18.47% ethanol, and 6.41 ± 1.00 gL^{-1} residual sugar [39]. Their study reported a starting sugar concentration of 30°Brix, which is higher than the present study, contributing to the different but proportional results. An analysis of commercial Amarone wines over four vintages (1998–2001) reported similar acetic acid levels of 0.52–0.62 g L^{-1}, ethanol levels of 15.15–15.88%, and residual sugar levels of 0.29–0.8%, equating to 2.9–8 g L^{-1} [6]. The wines in this current study that were made from partially dehydrated grapes had a starting sugar concentration of 28.0°Brix, resulting in an ethanol range of 14.7–15.3%, which is proportional to the starting sugar concentration of the Amarone wines outlined in the literature. Similarly, the high starting sugar wines fermented in this study with EC1118 had an acetic acid concentration of 0.36 g/L^{-1}; this is lower than the Amarone values reported in the literature, which is likely due to the lower starting sugar concentration, while CN1 produced even lower levels of acetic acid at 0.20 g/L^{-1}. This result suggests the potential commercial application of the CN1 yeast to winemaking using partially dehydrated grapes to assist in mitigating the quality challenges associated with undesirable oxidation compounds in the final wine [10,13]. This wine style in Ontario in commercial production targets starting sugar concentrations of the dried fruit between 27–28°Brix. Although CN1 did not ferment the 28°Brix must to complete dryness, Amarone wines are also found with residual sugar [6,39]. Additionally, Alessandria et al. [40] found that autochthonous yeast yielded incomplete sugar transformation, but the authors suggest that this result should not be considered negative for this type of wine, as residual sugar is typical for some wines made from partially dehydrated grapes, offering an opportunity for stylistic considerations for the winemaker [39–42]. Further, studies are currently underway to evaluate the sugar range over which CN1 does ferment to dryness.

Despite the lower production of acetic acid by CN1 in the wines, this yeast produced higher concentrations of glycerol in comparison to EC1118. It has been well-established that glycerol is produced as an intracellular osmolyte in *S. cerevisiae* under hyperosmotic stress during wine fermentations accompanied by acetic acid production. The link between these two metabolites in *S. cerevisiae* under hyperosmotic stress is based on a redox balance of the NAD$^+$/NADH system. The formation of glycerol generates NAD$^+$ [15,17,43]. Acetic acid production from acetaldehyde reduces NAD$^+$ to NADH through the activity of a NAD$^+$-dependent aldehyde dehydrogenase, and corrects the redox shift [17,18,44]. We recently reported a 24-fold higher NAD$^+$/total NAD(H) ratio in *S. cerevisiae* on fermentation day 2 during fermentation of 39°Brix juice compared to 20°Brix juice, which was correlated with higher glycerol production followed by acetic acid production [17]. In this current study, higher acetic acid production under osmotic stress was also noted in both yeast strains at the 28°Brix treatment compared to the 23°Brix treatment. However, *S. bayanus* CN1 produced more glycerol, but less acetic acid, in comparison to *S. cerevisiae* EC1118 at this higher brix condition. *S. bayanus* CN1 has a different response to osmotic stress than *S. cerevisiae*. Acetic acid may still be produced by *S. bayanus* as a response to glycerol production, but it may be further metabolized within the yeast as opposed to being released from the cell into the wine. Alternatively, a different metabolite may be used to reduce NAD$^+$ to NADH for redox balance, resulting in the lower acetic acid in the wine. Additional studies investigating the NAD(H) ratios in CN1 and yeast metabolites will provide insight on the mechanism and regulation of acetic acid production in high sugar fermentations in this yeast.

Wine color provides a quick reference of potential quality for consumers. The consumer can gather information about the wine's age, condition, body, and possible defects simply by looking at the wine as it leaves the bottle [45]. The basis for red wine color is anthocyanin content, and major secondary factors that are known to affect color density are pH and sulfur dioxide (SO$_2$) content. Interestingly, despite the low pigment content present in the wines vinified by *S. bayanus* CN1, at both sugar levels, they displayed a higher percentage of red-colored pigments than wines produced by EC1118. The CN1 control wine also showed a lower wine color hue compared to the other treatments. This could be caused by the lower pH in wines vinified by *S. bayanus*. It is accepted in the literature that the structure and color of anthocyanins are affected by pH, as acidification enhances the color intensity of red wine via the formation of the flavylium cation [46]. In addition to their direct role on color, anthocyanins can also contribute to the taste and chemical characteristics of wine because of their interactions with other molecules [47,48]. Therefore, they could have influenced the sensorially perceptible differences in the wines that were detected in this study. This is in agreement with the existing literature that found perceptible sensorial differences between the Amarone wines fermented with commercial *S. cerevisiae* yeast and those fermented with the inclusion of autochthonous yeast and non-*S. cerevisiae* yeast [39,49]. It is also important to note that there are differences in the wines in other categories; namely, orthonasal and/or retronasal sensory differences, as well as discrepancies in ethanol or residual sugar concentrations that could contribute to discriminating among the wines. The desirable higher percentage of red-colored pigments associated with CN1 and the established sensory differences amongst the treatments raise further questions about the organoleptic implications of using this yeast for wine production from partially dehydrated grapes. The differences are yet to be fully characterized; approaches such as quantitative sensory profiling and consumer preference testing would be useful in this regard.

5. Conclusions

This study lays the groundwork for further investigation of the potential of *S. bayanus* CN1 yeast for winemaking from partially dehydrated grapes in Ontario and other geographic regions that experience cool or marginal climates for grape growing. Although vinifying grapes for Vin de Curé poses risks for winemakers of increased oxidative compounds, the reward is in a high-value product that also adds diversity to the portfolio of a winery as well as its region. The findings on the isolate CN1 reported in this study are positive with respect to the legislated limits on oxidative compounds and desired red color hue, and have established sensory differences from the accepted commercial standard EC1118. Further to that, we recommend an additional sensory evaluation of wine made from *S. bayanus* CN1 in order to more fully understand its market potential. Additionally, understanding the difference between glycerol and acetic acid production of CN1 in comparison to *S. cerevisiae* EC1118 might contribute to the management of high acetic acid frequently associated with high sugar fermentations.

Author Contributions: D.L.I. and G.P. conceived and designed the experiments; L.D. and F.D.P. conducted the fermentations; L.D. performed the chemical analysis; J.K. conducted the color and sensory analysis; F.Y. prepared the yeast culture for fermentations; A.B., F.Y. and J.K. performed yeast identification; C.N. isolated the local *S. bayanus* strain. J.K., F.Y., G.P. and D.L.I. contributed to the writing of the manuscript.

funding: This project was funded by an Ontario Research Fund–Research Excellence grant (ORF RE-05-038) and a grant from the Natural Sciences and Engineering Research Council of Canada (NSERC 238872-2012).

Acknowledgments: We would like to thank Pillitteri Estates Winery for the donation of the grapes, and Cave Spring Cellars of the use of their facility for drying of the grapes.

Conflicts of Interest: The authors declare no conflict of interest.

References

1. Aylward, D.K. A Documentary of Innovation Support among New World Wine Industries. *J. Wine Res.* **2003**, *14*, 31–43. [CrossRef]
2. Frank, A.; Eyler, R. The Economic Impact of the Wine and Grape Industry in Canada 2015. Available online: http://www.canadianvintners.com/wp-content/uploads/2017/06/Canada-Economic-Impact-Report-2015.pdf (accessed on 11 September 2018).
3. Shaw, A.B. The Niagara Peninsula viticultural area: A climatic analysis of Canada's largest wine region. *J. Wine Res.* **2005**, *16*, 85–103. [CrossRef]
4. Shaw, T.B. Climate change and the evolution of the Ontario cool climate wine regions in Canada. *J. Wine Res.* **2017**, *28*, 13–45. [CrossRef]
5. Cyr, D.; Kusy, M.; Shaw, A.B. Climate change and the potential use of weather derivatives to hedge vineyard harvest rainfall risk in the Niagara region. *J. Wine Res.* **2010**, *21*, 207–227. [CrossRef]

6. Pagliarini, E.; Tomaselli, N.; Brenna, O.V. Study on sensory and composition changes in Italian Amarone Valpolicella red wine during aging. *J. Sens. Stud.* **2004**, *19*, 422–432. [CrossRef]

7. Figueiredo-González, M.; Cancho-Grande, B.; Simal-Gándara, J. Effects on colour and phenolic composition of sugar concentration processes in dried-on- or dried-off-vine grapes and their aged or not natural sweet wines. *Trends Food Sci. Technol.* **2013**, *31*, 36–54. [CrossRef]

8. Frangipane, M.T.; Torresi, S.; Santis, D.D.; Massantini, R. Effect of drying process in chamber at controlled temperature on the grape phenolic compounds. *Ital. J. Food Sci.* **2012**, *24*, 1–7.

9. Bellincontro, A.; Matarese, F.; D'Onofrio, C.; Accordini, D.; Tosi, E.; Mencarelli, F. Management of postharvest grape withering to optimise the aroma of the final wine: A case study on Amarone. *Food Chem.* **2016**, *213*, 378–387. [CrossRef] [PubMed]

10. Bellincontro, A.; De Santis, D.; Botondi, R.; Villa, I.; Mencarelli, F. Different postharvest dehydration rates affect quality characteristics and volatile compounds of Malvasia, Trebbiano and Sangiovese grapes for wine production. *J. Sci. Food Agric.* **2004**, *84*, 1791–1800. [CrossRef]

11. Ontario Regulation 406/00 Rules of Vintners Quality Alliance Ontario Relating to Terms for VQA Wine. Available online: https://www.ontario.ca/laws/regulation/000406/v27 (accessed on 7 September 2018).

12. Heit, C. Acetic acid and ethyl acetate production during high Brix fermentations: Effect of yeast strain. *Am. J. Enol. Vitic.* **2013**, *64*, 416A. [CrossRef]

13. Costantini, V.; Bellincontro, A.; De Santis, D.; Botondi, R.; Mencarelli, F. Metabolic changes of Malvasia grapes for wine production during postharvest drying. *J. Agric. Food Chem.* **2006**, *54*, 3334–3340. [CrossRef] [PubMed]

14. Cliff, M.A.; Pickering, G.J. Determination of odour detection thresholds for acetic acid and ethyl acetate in ice wine. *J. Wine Res.* **2006**, *17*, 45–52. [CrossRef]

15. Nevoigt, E.; Stahl, U. Osmoregulation and glycerol metabolism in the yeast *Saccharomyces cerevisiae*. *FEMS Microbiol. Rev.* **1997**, *21*, 231–241. [CrossRef] [PubMed]

16. Erasmus, D.J.; Cliff, M.; van Vuuren, H.J.J. Impact of yeast strain on the production of acetic acid, glycerol, and the sensory attributes of Icewine. *Am. J. Enol. Vitic.* **2004**, *55*, 371–387.

17. Yang, F.; Heit, C.; Inglis, D. Cytosolic redox status of wine yeast (*Saccharomyces cerevisiae*) under hyperosmotic stress during Icewine fermentation. *Fermentation* **2017**, *3*, 61. [CrossRef]

18. Pigeau, G.M.; Inglis, D.L. Upregulation of ALD3 and GPD1 in *Saccharomyces cerevisiae* during Icewine fermentation. *J. Appl. Microbiol.* **2005**, *99*, 112–125. [CrossRef] [PubMed]

19. Pigeau, G.M.; Inglis, D.L. Response of wine yeast (*Saccharomyces cerevisiae*) aldehyde dehydrogenases to acetaldehyde stress during Icewine fermentation. *J. Appl. Microbiol.* **2007**, *103*, 1576–1586. [CrossRef] [PubMed]

20. Kontkanen, D.; Inglis, D.L.; Pickering, G.J.; Reynolds, A. Effect of yeast inoculation rate, acclimatization, and nutrient addition on Icewine fermentation. *Am. J. Enol. Vitic.* **2004**, *55*, 363–370.

21. Garofalo, C.; Berbegal, C.; Grieco, F.; Tufariello, M.; Spano, G.; Capozzi, V. Selection of indigenous yeast strains for the production of sparkling wines from native Apulian grape varieties. *Int. J. Food Microbiol.* **2018**, *285*, 7–17. [CrossRef] [PubMed]

22. Garafalo, C.; Khoury, M.El.; Lucas, P.; Bely, M.; Russo, P.; Spano, G.; Capozzi, V. Autochthonous starter cultures and indigenous grape variety for regional wine production. *J. Appl. Microbiol.* **2015**, *118*, 1395–1408. [CrossRef] [PubMed]

23. Capozzi, V.; Spano, G. Food microbial biodiversity and "microbes of protected origin". *Front. Microbiol.* **2011**, *2*, 237. [CrossRef] [PubMed]

24. Capozzi, V.; Russo, P.; Spano, G. Microbial information regimen in EU geographical indications. *World Pat. Inf.* **2012**, *34*, 229–231. [CrossRef]

25. Capozzi, V.; Garofalo, C.; Chiriatti, M.A.; Grieco, F.; Spano, G. Microbial terroir and food innovation: The case of yeast biodiversity in wine. *Microbiol. Res.* **2015**, *181*, 75–83. [CrossRef] [PubMed]

26. Rantsiou, S.; Campolongo, S.; Alessandria, V.; Rolle, L.; Torchio, F.; Cocolin, L. Yeast populations associated with grapes during withering and their fate during alcoholic fermentation of high-sugar must. *Aust. J. Grape Wine Res.* **2013**, *19*, 40–46. [CrossRef]

27. Nurgel, C.; Inglis, D.L.; Pickering, G.J.; Reynolds, A.; Brindle, I. Dynamics of indigenous and inoculated yeast populations in Vidal and Riesling Icewine fermentations. *Am. J. Enol. Vitic.* **2004**, *55*, 435A.

28. Eglinton, J.M.; McWilliam, S.J.; Fogarty, M.W.; Francis, I.L.; Kwiatkowski, M.J.; Hoj, P.B.; Henschke, P.A. The effect of *Saccharomyces bayanus*-mediated fermentation on the chemical composition and aroma profile of Chardonnary wine. *Aust. J. Grape Wine Res.* **2000**, *6*, 190–196. [CrossRef]

29. Yang, F. Study of New Yeast Strains as Novel Starter Cultures for Riesling Icewine Production. Master's Thesis, Brock University, St. Catharines, ON, Canada, November 2010.

30. Nurgel, C.; Pickering, G.J.; Inglis, D.L. Sensory and chemical characteristics of Canadian ice wines. *J. Sci. Food Agric.* **2004**, *84*, 1675–1684. [CrossRef]

31. Zoecklein, B.W.; Fugelsang, K.C.; Gump, B.; Nury, F.S. *Wine Analysis and Production*; Springer: New York, NY, USA, 1995; 621p.

32. Iland, P.; Bruer, N.; Edwards, G.; Caloghiris, S.; Cargill, M.; Wilkes, E.; Iiland, J. *Chemical Analysis of Grapes and Wine: Techniques and Concepts*, 2nd ed.; Patrick Wine Promotions: Athelstone, Australia, 2013; pp. 88–89.

33. Kemp, S.E.; Hollowood, T.; Hort, J. *Sensory Evaluation a Practical Handbook*, 1st ed.; Chichester Ames, Iowa; Wiley: Hoboken, NJ, USA, 2009.

34. Huang, C.H.; Lee, F.L.; Tai, C.J. The *β-tubulin* gene as a molecular phylogenetic marker for classification and discrimination of the Saccharomyces sensu stricto complex. *Anton. Leeuw.* **2009**, *95*, 135–142. [CrossRef] [PubMed]

35. González, S.S.; Barrio, E.; Gafner, J.; Querol, A. Natural hybrids from *Saccharomyces cerevisiae*, *Saccharomyces bayanus* and *Saccharomyces kudriavzevii* in wine fermentations. *FEMS Yeast Res.* **2006**, *6*, 1221–1234. [CrossRef] [PubMed]

36. Pérez-Través, L.; Lopes, C.A.; Querol, A.; Barrio, E. On the complexity of the *Saccharomyces bayanus* taxon: Hybridization and potential hybrid speciation. *PLoS ONE* **2014**, *9*, e93729. [CrossRef] [PubMed]

37. Hittinger, C.T. *Saccharomyces* diversity and evolution: A budding model genus. *Trends Genet.* **2013**, *29*, 309–317. [CrossRef] [PubMed]

38. Sulo, P.; Szabóová, D.; Bielik, P.; Poláková, S.; Šoltys, K.; Jatzová, K.; Szemes, T. The evolutionary history of *Saccharomyces* species inferred from completed mitochondrial genomes and revision in the 'yeast mitochondrial genetic code. *DNA Res.* **2017**, *24*, 571–583. [CrossRef] [PubMed]

39. Azzolini, M.; Tosi, E.; Faccio, S.; Lorenzini, M.; Torriani, S.; Zapparoli, G. Selection of *Botrytis cinerea* and *Saccharomyces cerevisiae* strains for the improvement and valorization of Italian passito style wines. *FEMS Yeast Res.* **2013**, *13*, 540–552. [CrossRef] [PubMed]

40. Alessandria, V.; Giacosa, S.; Campolongo, S.; Rolle, L.; Rantsiou, K.; Cocolin, L. Yeast population diversity on grape during on-vine withering and their dynamics in natural and inoculated fermentations in the production of icewines. *Food Res. Int.* **2013**, *54*, 139–147. [CrossRef]

41. Giordano, M.; Rolle, L.; Zeppa, G.; Gerbi, V. Chemical and volatile composition of three Italian sweet white passito wines. *OENO ONE* **2009**, *43*, 159–170. [CrossRef]

42. Urso, R.; Rantsiou, K.; Dolci, P.; Rolle, L.; Comi, G.; Cocolin, L. Yeast biodiversity and dynamics during sweet wine production as determined by molecular methods. *FEMS Yeast Res.* **2008**, *8*, 1053–1062. [CrossRef] [PubMed]

43. Papapetridis, I.; van Dijk, M.; van Maris, A.J.A.; Pronk, J.T. Metabolic engineering strategies for optimizing acetate reduction, ethanol yield and osmotolerance in *Saccharomyces cerevisiae*. *Biotechnol. Biofuels* **2017**, *10*, 107. [CrossRef] [PubMed]

44. Heit, C.; Martin, S.J.; Yang, F.; Inglis, D.L. Osmoadaptation of wine yeast (*Saccharomyces cerevisiae*) during Icewine fermentation leads to high levels of acetic acid. *J. Appl. Microbiol.* **2018**. [CrossRef] [PubMed]

45. Kilcast, D. *Instrumental Assessment of Food Sensory Quality: A Practical Guide*, 1st ed.; Woodhead Publishing: Cambridge, UK, 2013; pp. 1–658.

46. Somers, T.C.; Evans, M.E. Wine quality: Correlations with colour density and anthocyanin equilibria in a group of young red wines. *J Sci. Food Agric.* **1974**, *25*, 1369–1379. [CrossRef]

47. Mazza, G.; Fukumoto, L.; Delaquis, P.; Girard, B.; Ewert, B. Anthocyanins, phenolics, and color of Cabernet Franc, Merlot, and Pinot Noir Wines from British Columbia. *J. Agric. Food Chem.* **1999**, *47*, 4009–4017. [CrossRef] [PubMed]

48. Vidal, S.; Francis, L.; Noble, A.; Kwiatkowski, M.; Cheynier, V.; Waters, E. Taste and mouth-feel properties of different types of tannin-like polyphenolic compounds and anthocyanins in wine. *Anal. Chim. Acta* **2004**, *513*, 57–65. [CrossRef]

49. Azzolini, M.; Fedrizzi, B.; Tosi, E.; Finato, F.; Vagnoli, P.; Scrinzi, C.; Zapparoli, G. Effects of *Torulaspora delbrueckii* and *Saccharomyces cerevisiae* mixed cultures on fermentation and aroma of Amarone wine. *Eur. Food Res. Technol.* **2012**, *235*, 303–313. [CrossRef]

fermentation

MDPI

Review

Lachancea thermotolerans, the Non-*Saccharomyces* Yeast that Reduces the Volatile Acidity of Wines

Alice Vilela

Department of Biology and Environment, Enology Building, School of Life Sciences and Environment, Chemistry Research Centre of Vila Real (CQ-VR), University of Trás-os-Montes and Alto Douro (UTAD), 5000-801 Vila Real, Portugal; avimoura@utad.pt

Received: 7 June 2018; Accepted: 16 July 2018; Published: 19 July 2018

Abstract: To improve the quality of fermented drinks, or more specifically, wine, some strains of yeast have been isolated, tested and studied, such as *Saccharomyces* and non-*Saccharomyces*. Some non-conventional yeasts present good fermentative capacities and are able to ferment in quite undesirable conditions, such as the case of must, or wines that have a high concentration of acetic acid. One of those yeasts is *Lachancea thermotolerants* (*L. thermotolerans*), which has been studied for its use in wine due to its ability to decrease pH through L-lactic acid production, giving the wines a pleasant acidity. This review focuses on the recent discovery of an interesting feature of *L. thermotolerans*—namely, its ability to decrease wines' volatile acidity.

Keywords: pioneering winemaking techniques; peculiar yeasts; volatile acidity; fermented drinks

1. Introduction

The pioneering of winemaking techniques and new yeast strains contributes to improving the quality of wines worldwide and offering solutions to various problems, such as increased sugar concentrations at grape maturity, or excessively acidic wines. Some non-*Saccharomyces*, as well as some non-conventional species of *Saccharomyces*, present good fermentative capacities and, are able to produce wines with lower levels of ethanol and higher concentrations of glycerol [1]. They are also able to avoid stuck fermentations, as they can grow at lower temperatures [2,3] as well as being nitrogen [4] and salt tolerant [5]. Moreover, mixed inoculations of non-*Saccharomyces*, *S. cerevisiae* yeasts, and lactic acid bacteria (LAB) in sequential fermentations are of great interest to the wine industry for various technological and sensorial reasons [6]. In addition, a peculiar microbial footprint that is characteristic of a particular wine region may be imprinted onto a wine if inoculation with autochthonous yeast is performed [3].

A non-*Saccharomyces* species not yet well-explored with huge biotechnological potential is *Lachancea thermotolerans* [7], formerly known as *Kluyveromyces thermotolerans* [8]. The genus *Lachancea* was proposed by Kurtzman in 2003 to accommodate a group from several different genera showing similarities at the rRNA level. According to Lachance and Lachancea [9], the genus continues to anchorage 11 other species to this day: *L. cidri*, *L. dasiensis*, *L. fantastica*, *L. fermentati*, *L. kluyveri*, *L. lanzarotensis*, *L. meyersi*, *L. mirantina*, *L. nothofagi*, *L. quebecensis*, and *L. walti*. As so-called protoploid *Saccharomycetaceae*, the *Lachancea* species has diverged from the *S. cerevisiae* lineage prior to the ancestral whole genome duplication, and as such, offers a complementary model for studying evolution and speciation in yeast [10].

Another peculiarity of *L. thermotolerans* is its ability to produce L-lactic acid during alcoholic fermentation [11]. Although lactic acid production is uncommon among yeasts, it is of great biotechnological interest in regard to fermentation processes where alcoholic fermentation with concomitant acidification is a benefit, such as winemaking [12].

Sometimes, to select a certain strain with specific enological features, scientists choose to use a genetic engineering approach. However, these techniques are unfortunately quite time-consuming and expensive. Moreover, metabolic engineering based on recombinant technology has some regulatory issues, such as the use of genetically modified organisms (GMO) in the food and wine industry [13]. The alternative solution to genetic manipulation is evolutionary engineering [14,15], which allows for improvements to phenotypes of choice. This methodology is based on the combination of confined environmental selection and natural variability. Evolutionary engineering aims to create an improved strain based on the selection of behavioral differences between individual cells within a population. The reason why non-recombinant strategies based on evolutionary engineering are eye-catching is that they may be able to generate better-quality strains that are not considered to be GMOs. Evolutionary engineering has long been used for generating new industrial strains [16,17], and of course, the first step is always to be well-focused on the selection criteria.

Our starting point for this research was the following question: Are indigenous yeasts—or more specifically, *L. thermotolerans*—able to reduce volatile acidity from musts and wines?

2. The Vinegar Taint Problem in Wine

In excessive quantities, volatile acids are considered to be a spoilage characteristic of wines, as they confer an unpleasant vinegary aroma along with an acrid taste. The main component of the volatile acidity of wines and musts is acetic acid. The maximum acceptable limit for volatile acidity in most wines is 1.2 g L^{-1} of acetic acid [18], but the aroma threshold for acetic acid depends on the variety and style of the wine, being the vinegary smell recognizable at acetic acid concentrations of 0.8–0.90 g L^{-1} [19]. Acetic acid can be formed at any time during the wine-making process. It can appear on the grapes or the grape-must due to a myriad of yeasts (*Hansenula* spp. and *Brettanomyces bruxellensis*), filamentous fungi (*Aspergillus niger*, *Aspergillus tenuis*, *Cladosporium herbarum*, *Rhizopus arrhizus*, and *Penicillium* spp), and bacteria (LAB-like indigenous *Lactobacilli*, otherwise known as "ferocious", and acetic acid bacteria). It can also appear during the alcoholic fermentation process as a by-product of *S. cerevisiae* sugars' metabolism, or due to some contamination by spoilage yeasts (*Pichia anomala*, *Candida krusei*, *Candida stellate*, *Hansaniaspora uvarum/Kloeckera apiculate*, and *Saccharomycodes ludwigii*) or bacteria (*Acetobacter pasteurianus* and *Acetobacter liquefaciens* that survive during fermentation); after MLF (malolactic fermentation), due to heterofermentative species of *Oenococcus* and *Lactobacillus* that have the potential to produce acetic acid through the metabolism of residual glucose (usually no more than 0.1–0.2 g L^{-1} of acetic acid); or in the bottled wine, due to spoilage by contaminating yeasts and/or bacteria [20,21].

3. *L. thermotolerans*' Main Features in Alcoholic Drinks

In recent years, *Lachancea thermotolerans* (formerly *Kluyveromyces thermotolerans*) has been studied for its use in wine and beer, due to its ability to decrease pH through lactic acid production [22]. *K. thermotolerans* was alienated from the other species of *Kluyveromyces* and placed in *Lachancea* due to its distinct genetic and metabolic differences, as well as its genetic similarity to other members of the *Lachancea* genus [8].

This yeast is often found in a selection of fruits, such as on the surface of grapes. Consequently, it is present at the beginning of many fermentations before *Saccharomyces*' domination. Due to the production of L-lactic acid (from 0.23 to 9.6 g L^{-1}, depending on the different trial conditions [23,24]), wine produced by fermentation with *L. thermotolerans* is considered to display some different sensory properties—mainly in terms of mouth-feel—with an increased acidic taste [23–25]. Some winemakers desire to have varying degrees of these traits in their wines, and now, *L. thermotolerans* is present in a few commercial yeast inoculates.

For instance, strain 617 of *L. thermotolerans* was selected amongst other non-*Saccharomyces* yeasts to perform combined fermentations with *S. cerevisiae*, in order to increase the acidity and quality of Spanish Airén wine [25]. Although this Spanish grape variety is considered to be very neutral and productive, the wines it is used in are usually considered to be low quality due to its high sugar content and lack of acidity [25].

During wine fermentation, *L. thermotolerans* also causes an increase in levels of ethyl lactate [19]. Due to these metabolic features (lactic acid and ethyl lactate production), this yeast is currently being studied for the purposes of producing beer without the necessity of LAB inoculation [26]. Thus, using *L. thermotolerans* to produce beer with a sourer taste may be simpler than trying to maintain a co-fermentation with yeasts and bacteria.

However, the metabolic pathway of converting sugars to lactic acid by *L. thermotolerans* is not completely understood. Recent discoveries to date have shown that levels of lactic acid between 1–9 g L^{-1} were found in wine which was fermented with this yeast species [23]. The metabolism of sugars into lactic acid is also a way to reduce the level of alcohol in wines, and a reduction of up to 0.5 to 1% (v/v) of alcohol is possible [23].

Benito et al. [27] investigated the application of *L. thermotolerans* and *Schizosaccharomyces pombe* as an alternative to the classic malolactic fermentation in a wine made with *Vitis vinifera* L. cultivar Tempranillo. While *S. pombe* totally consumed the malic acid, *L. thermotolerans* produced lactic acid, which allowed excessive deacidification to be avoided. In addition, the results from the fermentation trails showed positive differences in several parameters, such as acetic acid, glycerol, acid profile, sensory evaluation, color, and anthocyanin profile. Moreover, Benito et al. [24] also demonstrated that wines crafted through this technique had biogenic amines (BAs) levels lower than 2 mg L^{-1}. The combined use of two non-*Saccharomyces* strains allowed for a reduction of the value of all measured BAs, in comparison with the use of *Saccharomyces* and malolactic fermentation, from 0.44 *vs.* 1.46 mg L^{-1} and 1.71 *vs.* 2.18 mg L^{-1}, respectively. The authors [24] also stated that these differences should be attributed to the ability of *Schizosaccharomyces pombe* to metabolize urea [28].

4. Strain Isolation and Wine Biodeacetification

Several approaches have been developed in which regards to the deacetification of wines, including "empirical" enological techniques, where acidic wines are refermented by mixing them with marc from a finished wine fermentation, or by mixing them with freshly-crushed grapes or musts. More modern techniques have been explained and studied in enological, biochemical, and microbiological terms [20,29–31]. Under aerobic conditions, acetate can be used as a sole source of carbon and energy for the purposes of energy generation and cellular biomass [32]. This feature is not just present in *S. cerevisiae* strains—some *Zygosaccharomyces bailii* strains also display biphasic growth in media containing mixtures of glucose and acetic acid [33].

In previous works, such as a study by Vilela et al. [34], several yeast strains have been isolated (e.g., *Saccharomyces* and non-*Saccharomyces*) in Wallerstein Laboratory Nutrient Agar (WL) media using the refermentation processes of acidic wines, at winery scale [34]. Among all isolates, a group of yeasts was selected for testing for their ability to consume acetic acid in the presence of glucose, using a differential medium containing acetic acid and glucose adapted from Schuller et al. [35], shown in Figure 1.

Four of those isolates in this medium were obtained and characterized by fingerprinting with primer T3B, then identified by the amplification of the D1–D2 variable domain at the 5' end of the 26S rDNA (nucleotides 63–642 for *S. cerevisiae*) with primers NL-1 and NL-4. The amplified fragments were subsequently sequenced. As shown in Figure 1, this method confirmed the presence of *L. thermotolarans*, coded in our work as strain number 44C [29].

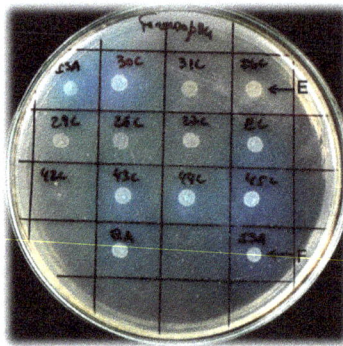

Figure 1. Growth and color change (due to pH changes) of the differential medium with 0.5% (v/v) acetic acid, 0.05% (w/v) glucose, and bromocresol green (0.005% (w/v)) at pH 4.0, indicating the simultaneous consumption of glucose and acetic acid by the isolated strains, namely "44C". E: *S. cerevisiae* PYCC4072 (negative control); F: *Z. bailii* ISA1307 (positive control). Retrieved from Vilela et al. [34].

Subsequently, the effect of glucose and acetic acid concentrations and aeration conditions, on the consumption of acetic acid by the previously mentioned strain, were studied at laboratory scale. The strain *Z. bailii* ISA1307 was used as a reference strain. The results showed that *L. thermotolerans* 44C was able to degrade 28.2% of the initial acid when grown under limited aerobic conditions in a mixed substrate medium which contained glucose (5.0%, w/v) and acetic acid (5.0 g L^{-1}). Moreover, strain 44C also presented the ability to degrade acetic acid in media with 5.0% or 0.75% (w/v) of glucose, under limited aerobic conditions. Although the higher initial concentration of glucose did not alter the rate of acetic acid consumption by strain 44C, this strain did decrease the rate of glucose consumption [29].

To verify the potential application of *L. thermotolerans* 44C in refermentation processes, the strain was inoculated in a mixed medium containing two-thirds of minimal medium [36], supplemented with one-third of an acidic white wine. Once again, *Z. bailii* ISA1307 was used as a control strain. The volatile acidity of the mixture was 1.13 g L^{-1} of acetic acid, corresponding to the values usually found in acidic wines. Two wine-supplemented mineral media (Table 1) were tested: The first medium simulated the refermentation of a wine with freshly crushed grapes or with grape-must [13% glucose (w/v), 4% ethanol (v/v)]; the second wine-supplemented mineral medium simulated the refermentation of a wine with the residual marc from a finished wine fermentation [3.3% glucose (w/v), 10% ethanol (v/v)]. Acetic acid consumption was again evaluated under aerobic or limited aerobic conditions to assess whether aeration was a limiting factor in the process. Once it was known that higher glucose concentration levels could lead to a decreased rate of glucose consumption, the rate of acetic acid and glucose consumption was evaluated at the end of 48 and 72 h for the first and second medium, respectively [29].

Strains *Z. bailii* ISA1307 and *L. thermotolerans* 44C were efficient in terms of acetic acid consumption in the high-glucose medium and aerobic conditions, as 94.8 and 94.6% of the initial acetic acid was consumed (Table 1). However, the efficiency of *L. thermotolerans* 44C in acetic acid consumption decreased significantly in the high-glucose concentration medium under limited aerobic conditions, as only 15.3% of the initial acetic acid was consumed [29], as shown in Table 1.

Table 1. Percentage of acetic acid (*italic letters*) and glucose (**bold letters**) consumption after refermentation of wine-supplemented culture medium, containing glucose 13% (*w/v*) and ethanol 4% (*v/v*) or glucose 3.3% (*w/v*) and ethanol 10% (*v/v*), after 48 and 72 h of incubation, respectively. Results obtained for strains and culture conditions with the same letter are not significantly different (*p* < 0.001) [29].

Yeast strains	Glucose (13%, *w/v*) Ethanol (4%, *v/v*)		Glucose (3.3%, *w/v*) Ethanol (10%, *v/v*)	
	Aerobic Conditions	Limited Aerobic Conditions	Aerobic Conditions	Limited Aerobic Conditions
	Acetic acid **Glucose**	*Acetic acid* **Glucose**	*Acetic acid* **Glucose**	*Acetic acid* **Glucose**
Z. bailii ISA 1307	*94.8 ± 3.30 c* **52.4 ± 2.62 c**	*40.9 ± 9.80 a* **38.8 ± 6.36 b**	*71.2 ± 3.02 b* **23.1 ± 5.60 a**	*41.6 ± 2.64 a* **39.4 ± 2.10 b**
L. thermotolerans 44C	*94.6 ± 4.79 d* **58.5 ± 8.60 c**	*15.25 ± 3.30 a* **31.0 ± 5.69 b**	*28.1 ± 1.70 c* **16.4 ± 1.76 a**	*17.4 ± 7.16 b* **30.4 ± 5.79 b**

5. Conclusions

Climate change has caused increased temperatures for many wine-growing regions around the world. *L. thermotolerans* offers a unique potential to counter this effect of global warming on wine grapes by producing acid during fermentation, which can moderately reduce alcohol levels and produce, also, high concentrations of the fruity-like flavor compound, ethyl lactate (using lactate as a precursor).

L. thermotolarans can also be used to develop a controlled biological deacetification process of wines with high volatile acidity. However, the ability of *L. thermotolerans* 44C to consume acetic acid is a highly oxygen-dependent one, which means that its metabolism must shift more towards respiration than to fermentation. The high amount of sugars present in the grape-must may also inhibit or delay *L. thermotolerans'* acetic acid consumption during refermentation processes.

Consequently, further research is needed before the deacetification of wines will be able to be added to the list of *Lanchancea's* enological features.

funding: We appreciate the financial support provided to the Research Unit (CQ-VR) in Vila Real (PEst-OE/QUI/UI0616/2014) by FCT—Portugal and COMPETE.

Conflicts of Interest: The author declares no conflict of interest.

References

1. Ciani, M.; Morales, P.; Comitini, F.; Tronchoni, J.; Canonico, L.; Curiel, J.A.; Gonzalez, R. Non-conventional Yeast Species for Lowering Ethanol Content of Wines. *Front. Microbiol.* **2016**, *7*, 642. [CrossRef] [PubMed]
2. Padilla, B.; Gil, J.V.; Manzanares, P. Past, and Future of Non-*Saccharomyces* Yeasts: From Spoilage Microorganisms to Biotechnological Tools for Improving Wine Aroma Complexity. *Front. Microbiol.* **2016**, *7*, 411. [CrossRef] [PubMed]
3. Lleixà, J.; Manzano, M.; Mas, A.; Portillo, M.C. *Saccharomyces* and non-*Saccharomyces* competition during microvinification under different sugar and nitrogen conditions. *Front. Microbiol.* **2016**, *7*, 1959. [CrossRef] [PubMed]
4. Brice, C.; Cubillos, F.A.; Dequin, S.; Camarasa, C.; Martínez, C. Adaptability of the *Saccharomyces cerevisiae* yeasts to wine fermentation conditions relies on their strong ability to consume nitrogen. *PLoS ONE* **2018**, *13*, e0192383. [CrossRef] [PubMed]
5. Dibalova-Culakova, H.; Alonso-del-Real, J.; Querol, A.; Sychrova, H. Expression of heterologous transporters in *Saccharomyces kudriavzevii*: A strategy for improving yeast salt tolerance and fermentation performance. *Int. J. Food Microbiol.* **2018**, *268*, 27–34. [CrossRef] [PubMed]

6. Minnaar, P.P.; Plessis, H.W.; du Paulsen, V.; Ntushelo, N.; Jolly, N.P.; du Toit, M. *Saccharomyces cerevisiae*, non-*Saccharomyces* yeasts and lactic acid bacteria in sequential fermentations: Effect on phenolics and sensory attributes of South African Syrah Wines. *S. Afr. J. Enol. Vitic.* **2017**, *38*, 237–244. [CrossRef]
7. Hranilovic, A.; Bely, M.; Masneuf-Pomarede, I.; Jiranek, V.; Albertin, W. The evolution of *Lachancea thermotolerans* is driven by geographical determination, anthropisation and flux between different ecosystems. *PLoS ONE* **2017**, *12*, e0184652. [CrossRef] [PubMed]
8. Kurtzman, C.P. Phylogenetic circumscription of *Saccharomyces, Kluyveromyces* and other members of the *Saccharomycetaceae*, and the proposal of the new genera *Lachancea, Nakaseomyces, Naumovia, Vanderwaltozyma*, and *Zygotorulaspora*. *FEMS Yeast Res.* **2003**, *4*, 233–245. [CrossRef]
9. Lachance, M.A.; Lachancea, K. *The Yeasts, a Taxonomic Study*; Kurtzman, C., Fell, J.W., Boekhout, T., Eds.; Elsevier: London, UK, 2011; pp. 511–519.
10. Souciet, J.L.; Dujon, B.; Gaillardin, C.; Johnston, M.; Baret, P.V.; Cliften, P.; Sherman, D.J.; Weissenbach, J.; Westhof, E.; Wincker, P.; et al. Comparative genomics of protoploid *Saccharomycetaceae*. *Genome Res.* **2009**, *19*, 1696–1709. [CrossRef] [PubMed]
11. Jolly, N.P.; Varela, C.; Pretorius, I.S. Not your ordinary yeast: Non-*Saccharomyces* yeasts in wine production uncovered. *FEMS Yeast Res.* **2014**, *14*, 215–237. [CrossRef] [PubMed]
12. Dequin, S.; Barre, P. Mixed lactic acid–alcoholic fermentation by *Saccharomyces cerevisiae* expressing the *Lactobacillus casei* L (+)–LDH. *Nat. Biotechnol.* **1994**, *12*, 173–177. [CrossRef]
13. Çakar, Z.P.; Turanli-Yildiz, B.; Alkim, C.; Yilmaz, U. Evolutionary engineering of *Saccharomyces cerevisiae* for improved industrially important properties. *FEMS Yeast Res.* **2012**, *12*, 171–182. [CrossRef] [PubMed]
14. Vanee, N.; Fisher, A.B.; Fong, S.S. Evolutionary Engineering for Industrial Microbiology. In *Reprogramming Microbial Metabolic Pathways. Subcellular Biochemistry*; Wang, X., Chen, J., Quinn, P., Eds.; Springer: Dordrecht, The Netherlands, 2012; Volume 64.
15. Fong, S.S. Evolutionary engineering of industrially important microbial phenotypes. In *The Metabolic Pathway Engineering Handbook: Tools and Applications*; Smolke, C.D., Ed.; CRC Press: New York, NY, USA, 2010; ISBN 978-142-0077-65-0.
16. Sonderegger, M.; Sauer, U. Evolutionary engineering of *Saccharomyces cerevisiae* for anaerobic growth on xylose. *Appl. Environ. Microbiol.* **2003**, *69*, 1990–1998. [CrossRef] [PubMed]
17. López-Malo, M.; García-Rios, E.; Melgar, B.; Sanchez, M.R.; Dunham, M.J.; Guillamón, J.M. Evolutionary engineering of a wine yeast strain revealed a key role of inositol and mannoprotein metabolism during low-temperature fermentation. *BMC Genom.* **2015**, *16*, 537. [CrossRef] [PubMed]
18. Office Internationale de la Vigne et du Vin. *International Code of Oenological Practices*; OIV: Paris, France, 2010.
19. Ribéreau-Gayon, P.; Glories, Y.; Maujean, A.; Dubourdieu, D. Alcohols, and other volatile compounds. The chemistry of wine stabilization and treatments. In *Handbook of Enology*, 2nd ed.; John Wiley & Sons Ltd.: Chichester, UK, 2006; Volume 2, pp. 51–64. [CrossRef]
20. Vilela-Moura, A.; Schuller, D.; Mendes-Faia, A.; Silva, R.F.; Chaves, S.R.; Sousa, M.J.; Côrte-Real, M. The impact of acetate metabolism on yeast fermentative performance and wine quality: Reduction of volatile acidity of grape-musts and wines—Minireview. *Appl. Microbiol. Biotechnol.* **2011**, *89*, 271–280. [CrossRef] [PubMed]
21. Cosme, F.; Vilela, A.; Filipe-Ribeiro, L.; Inês, A.; Nunes, F.-M. Wine microbial spoilage: Advances in defects remediation. In *Microbial Contamination and Food Degradation, Handbook of Bioengineering*, 1st ed.; Grumezescu, A., Holban, A.M., Eds.; Elsevier: Amsterdam, The Netherlands; Academic Press: New York, NY, USA, 2017; Volume 10, pp. 271–314. Available online: https://www.elsevier.com/books/microbial-contamination-and-food-degradation/grumezescu/978-0-12-811262-5 (accessed on 20 May 2018).
22. Hill, A. Traditional methods of detection and identification of brewery spoilage organisms. In *Brewing Microbiology: Managing Microbes, Ensuring Quality and Valorising Waste*; Series in Food Science, Technology and Nutrition; Woodhead: London, UK, 2015; Volume 289, 506p.
23. Gobbi, M.; Comitini, F.; Domizio, P.; Romani, C.; Lencioni, L.; Mannazzu, I.; Ciani, M. *Lachancea thermotolerans* and *Saccharomyces cerevisiae* in simultaneous and sequential co-fermentation: A strategy to enhance acidity and improve the overall quality of wine. *Food Microbiol.* **2013**, *33*, 271–281. [CrossRef] [PubMed]
24. Benito, S.; Hofmann, T.; Laier, M.; Lochbühler, B.; Schüttler, A.; Ebert, K.; Fritsch, S.; Röcker, J.; Rauhut, D. Effect on quality and composition of Riesling wines fermented by sequential inoculation with non-*Saccharomyces* and *Saccharomyces cerevisiae*. *Eur. Food Res. Technol.* **2015**, *241*, 707–717. [CrossRef]

25. Benito, Á.; Calderón, F.; Palomero, F.; Benito, S. Quality and Composition of Airén Wines Fermented by Sequential Inoculation of *Lachancea thermotolerans* and *Saccharomyces cerevisiae*. *Food Technol. Biotechnol.* **2016**, *54*, 135–144. [CrossRef] [PubMed]

26. Domizio, P.; House, J.F.; Joseph, C.M.L.; Bisson, L.F.; Bamforth, C.W. *Lachancea thermotolerans* as an alternative yeast for the production of beer. *J. Inst. Brew.* **2016**, *122*, 599–604. [CrossRef]

27. Benito, Á.; Calderón, F.; Benito, S. The Combined Use of *Schizosaccharomyces pombe* and *Lachancea thermotolerans*—Effect on the Anthocyanin Wine Composition. *Molecules* **2017**, *22*, 739. [CrossRef] [PubMed]

28. Lubbers, M.W.; Rodriguez, S.B.; Honey, N.K.; Thornton, R.J. Purification, and characterization of urease from *Schizosaccharomyces pombe*. *Can. J. Microbiol.* **1996**, *42*, 132–140. [CrossRef] [PubMed]

29. Vilela-Moura, A.; Schuller, D.; Mendes-Faia, A.; Côrte-Real, M. Reduction of volatile acidity of wines by selected yeast strains. *Appl. Microbiol. Biotechnol.* **2008**, *80*, 881–890. [CrossRef] [PubMed]

30. Vilela-Moura, A.; Schuller, D.; Falco, V.; Mendes-Faia, A.; Côrte-Real, M. Effect of refermentation conditions and micro-oxygenation on the reduction of volatile acidity by commercial *S. cerevisiae* strains and their impact on the aromatic profile of wines. *Int. J. Food Microbiol.* **2010**, *141*, 165–172. [CrossRef] [PubMed]

31. Vilela-Moura, A.; Schuller, D.; Mendes-Faia, A.; Côrte-Real, M. Effects of acetic acid, ethanol and SO_2 on the removal of volatile acidity from acidic wines by two *Saccharomyces cerevisiae* commercial strains. *Appl. Microbiol. Biotechnol.* **2010**, *87*, 1317–1326. [CrossRef] [PubMed]

32. Schüller, H.J. Transcriptional control of non-fermentative metabolism in the yeast *Saccharomyces cerevisiae*. *Curr. Genet.* **2003**, *43*, 139–160. [CrossRef] [PubMed]

33. Sousa, M.J.; Rodrigues, F.; Côrte-Real, M.; Leão, C. Mechanisms underlying the transport and intracellular metabolism of acetic acid in the presence of glucose in the yeast *Zygosaccharomyces bailii*. *Microbiology* **1998**, *144*, 665–670. [CrossRef] [PubMed]

34. Vilela, A.; Amaral, C.; Schuller, D.; Mendes-Faia, A.; Corte-Real, M. Combined use of Wallerstein and *Zygosaccharomyces bailii* modified differential media to isolate yeasts for the controlled reduction of volatile acidity of grape musts and wines. *J. Biotech Res.* **2015**, *6*, 43–53.

35. Schuller, D.; Côrte-Real, M.; Leão, C. A differential medium for the enumeration of the spoilage yeast *Zygosaccharomyces bailii* in wine. *J. Food Prot.* **2000**, *63*, 1570–1575. [CrossRef] [PubMed]

36. Van Uden, N. Transport-limited fermentation, and growth of *Saccharomyces cerevisiae* and its competitive inhibition. *Arch. Mikrobiol.* **1967**, *58*, 155–168. [CrossRef] [PubMed]

fermentation

MDPI

Article

Agronomical and Chemical Effects of the Timing of Cluster Thinning on Pinot Noir (Clone 115) Grapes and Wines

Paul F. W. Mawdsley, Jean Catherine Dodson Peterson and L. Federico Casassa *

Wine and Viticulture Department, California Polytechnic State University, San Luis Obispo, CA 93407, USA; pmawdsle@calpoly.edu (P.F.W.M.); jdodsonp@calpoly.edu (J.C.D.P.)
* Correspondence: lcasassa@calpoly.edu; Tel.: +1-805-756-2751

Received: 16 June 2018; Accepted: 26 July 2018; Published: 31 July 2018

Abstract: A two-year study was performed to evaluate the effects of the timing of cluster thinning on Pinot noir grapes and wines in the central coast of California. Vines were thinned to one cluster per shoot at three selected time-points during the growing season, and fruit was harvested and made into wine. No consistent effect of cluster thinning was found in wine phenolic profile or color across a cool (2016) and a warm (2017) growing season. The growing season had a more significant effect than the cluster thinning treatment for most parameters measured. There was no detectable overall sensory difference between the non-thinned control wines and any of the thinned treatment wines. Based on current results, Pinot noir vineyards on the central coast of California can support crop loads that result in Ravaz Index values from 3 to 6 without concern for impacting ripening potential or negatively affecting fruit composition.

Keywords: cluster thinning; yield manipulation; vine balance; crop load; Pinot noir; Central Coast of California

1. Introduction

Pinot noir (*Vitis vinifera* L.) is a challenging grape cultivar from both a viticultural and winemaking perspective. Viticulturally, Pinot noir grapevines produce compact clusters of thin-skinned berries, which increase susceptibility to fungal pathogens relative to other *V. vinifera* cultivars. Pinot noir grapes (and their wines) are also inherently low in phenols [1,2]. Phenols are biomolecules originally present in the grapes (and subsequently extracted into wine). Phenols can be broadly classified as simple phenols having a C6–C1 or C6–C3 structure and a single aromatic ring containing one or more hydroxyl groups; and polyphenols, which contain multiple phenol rings and are defined by a C6–C3–C6 structure bearing hydroxyl and non-hydroxyl substitutions [3]. In wines, polyphenols are responsible for color [4–6], tactile sensations such as astringency [7,8], and taste sensations such as bitterness [9–11]. In addition to sensory effects on astringency, taste, and aroma modulation [12], flavonoids also play a critical role in the chemical stability of the wine during aging as these molecules intervene in metal-catalyzed oxidation reactions [13,14]. Because of the relatively lower phenolic content of Pinot noir, wines produced from it are lighter in color and astringency than wines made from other cultivars [15]. Pinot noir is also notable for lacking acylated anthocyanins [2], which are abundant in other cultivars such as Cabernet Sauvignon or Syrah and may in turn provide more stable color [16,17]. As color is one of the main drivers of perceived wine quality [18], viticultural practices such as cluster thinning are often applied to Pinot noir grapes in an attempt to lower yields and influence fruit polyphenol composition by lowering vine crop load [19–25].

Polyphenols such as anthocyanins and tannins, and their reaction products, known as polymeric pigments [8], are positively associated with wine quality [18]. In turn, and as mentioned above, vineyard crop load manipulation techniques such as cluster thinning are often applied to influence phenolic development [19–25]. The traditional yield to fruit quality paradigm of a linear relationship with quality increasing as yield decreases [24–27], has been shown to be an oversimplification, and yields are more accurately described as a function of vine balance [28,29]. Vine balance, better described as the source/sink ratio, relates vine vegetative and reproductive growth, either through leaf area/yield (LA/Y) ratios [30] or through the ratio between dormant vine pruning weights and yields, the latter known as the Ravaz Index [31,32]. For most cultivars, in warm climates, 0.8–1.2 m^2 of leaf area is needed to ripen 1 kg of fruit, which generally results in a yield/pruning weight ratio of 5 to 10 [30].

Crop load metrics are dependent upon vine capacity, which is in turn influenced by regional and viticultural factors such as climate [33], canopy training and trellising [34], rootstock, and cultivar [21,35]. Grapes grown in cool climates require higher source/sink ratios than those grown in warmer regions because of lower daytime temperatures that restrict both leaf photosynthetic capacity and berry carbon assimilation [29,34,36]. As a result of this restricted photosynthetic capacity, the ability of grapes grown in cool climates to ripen fruit to commercially viable total soluble solids (TSS) levels is limited and is significantly affected by seasonal variations in weather [33,34]. As such, in cool climates, seasonal variations in weather may have a greater effect than the source/sink ratio on Pinot noir ripening [33].

Despite the prevalence of cluster thinning in Pinot noir, few studies have been conducted to investigate the effects of this viticultural technique on this cultivar. Indeed, most studies on cluster thinning have been conducted in warm, arid climates and on cultivars such as Cabernet Sauvignon [37,38] and Tempranillo [22,24,39]. The bulk of this research suggests conflicting results, whereby it has not been conclusively shown that manipulating yields by cluster thinning will uniformly affect fruit composition. For example, some research has indicated that cluster thinning can positively impact fruit composition [26,27,40], while other studies have found no effect on fruit composition [21,38,41,42] or that the effect of cluster thinning was dependent upon the climate conditions of any single growing season [20,21,24]. In other instances, compositional effects as a result of cluster thinning have been found in grapes, but these have not translated into the finished wines [27]. Pinot noir grown in cool climates may be more suited to benefit from cluster thinning because of the inherently low polyphenol content of the fruit and reduced carbon assimilation capacity of vines in cool climates [29,34,36].

The timing of cluster thinning may also have an impact on vine physiology and fruit composition. For example, it has been hypothesized that removing crop at bloom may lead to lower leaf transpiration rates, and therefore lower leaf photosynthesis rates [43], which could negate the desired effect of enhanced berry ripeness [21]. In addition, if photosynthesis rates remain unchanged but the source/sink ratio increases upon cluster thinning, the increased photo-assimilates may stimulate vegetative growth, counteracting or negating the benefits of the decreased crop load [29,32]. In a study spanning five seasons, early thinning at bloom increased berry weight in Cabernet Sauvignon, Riesling, and Chenin blanc, while late thinning performed at véraison was intermediate to early thinning and non-thinned vines [21]. However, this effect was not found in all years of the previous study. Cluster thinning applied at bloom to Pinot noir vines resulted in increased berry size [19], although there was no late thinning treatment included in the study. Other research has shown no effect of thinning on berry size in Tempranillo [22], thereby suggesting that there is likely a cultivar-specific response of berry size as a result of cluster thinning. Berry size reduction has traditionally been considered desirable from a winemaking perspective based on the empirical assumption that comparatively smaller berries have higher berry surface area/volume than larger ones. However, the relationship between berry size and phenolic composition is not a simple linear relationship, as berry skin and seed mass grow along with berry size [44], and therefore larger berries may not be necessarily undesirable as once thought. Multiple studies conducted with a variety of cultivars investigating the relationship between cluster thinning and berry size have found conflicting results dependent upon cultivar and growing season [19,21,22], and as such, there is no current conclusive understanding of the relationship between cluster thinning and berry size, and the subsequent effect of the latter on wine composition.

In the central coast of California (USA), consisting of Santa Barbara, San Luis Obispo, and Monterey counties, there are over 7000 hectares of Pinot noir being grown [45], representing a substantial contribution to the wine industry of the region. Indeed, in 2016, wine grapes were the most valuable crop produced in San Luis Obispo County [46]. Despite the economic importance of Pinot noir on the central coast of California, no research has been undertaken to understand the relationship between Pinot noir crop load and fruit quality in the cool climate of San Luis Obispo county of the central coast. While research in cooler areas such as Oregon's Willamette Valley (USA) have indicated that grape ripeness increased in a curvilinear fashion with increasing LA/Y ratios up to 1.25–1.75 m^2/kg [34], which is higher than the LA/Y ratios observed in warm climates [30], these source/sink ratios are intrinsically tied to the seasonal limitations of the region that result in inconsistent ripening and may not be translatable to more moderate cool climates such as California's central coast.

In the present study, the effect of crop load reduction by cluster thinning was explored for the first time on Pinot noir grapes (clone 115) and wines grown in the cool climate conditions of the Edna Valley of the Central Coast (San Luis Obispo County) of California (USA). The objectives of this study were to evaluate the effect of the timing of cluster thinning crop reduction on vine capacity, berry composition, and wine composition over two consecutive seasons. An additional objective was to identify appropriate crop loads in cool climate Pinot noir grown on the central coast of California.

2. Materials and Methods

2.1. Vineyard Site

This study was conducted at a commercial vineyard located in Edna Valley (35°11′58.3″ N 120°34′12.6″ W), San Luis Obispo County, California, during the 2016 and 2017 growing seasons. Treatments of cluster thinning were applied to Pinot noir grapevines (clone 115) planted in 1996 on 5C rootstock. Cluster thinning was applied to these Pinot noir grapevines at selected phenological growth stages. Cluster thinning was conducted by removing the second cluster from each fruiting shoot of the vine. Any third clusters (second crop) were also removed during the treatment, resulting in one cluster per fruiting shoot. No second crop thinning or reduction was performed in the control vines. Vineyard treatments were applied as 100-vine sets replicated five times (n = 5), organized as a randomized complete block. Bloom was defined as stage 23 on the modified Eichhorn and Lorenz scale [47], and occurred on 1 June in 2016 and 15 May in 2017. Cluster thinning was applied at four weeks post-bloom (bloom + 4) approximating fruit set, eight weeks post-bloom (bloom + 8) approximating véraison, and 12 weeks post bloom (bloom + 12) shortly before harvest. Vines were pruned to two ten bud canes trained in a vertical shoot positioning (VSP) system with two catch-wires. Canes removed during the 2018 winter pruning were collected and weighed on a per-vine basis to determine 2017 growing season vegetative growth. Vine spacing was 2.75 × 1.52 m in north–south aligned rows planted in silty clay loam soil. Precipitation and daily minimum and maximum temperatures were recorded from California Irrigation Information Management System (CIMIS) weather station 52 (35°18′19.6″ N 120°39′42.4″ W), located 14.41 km from the experimental site. Cumulative growing degree days (GDD) were calculated using a baseline temperature of 10 °C and the average daily temperature from 1 April to 31 October of each year [48].

2.2. Winemaking

Fruit was harvested when a composite sample of all treatments (n = 25, 250 berries each) reached 22.5 Brix. Harvest dates were 6 September 2016 and 6 September 2017. Fruit was harvested manually from three independent vineyard replications of each treatment (n = 3). Approximately 80 kg of fruit per replicate was harvested both in 2016 and 2017, for a total of 960 kg of fruit harvested in each season. The three replicates of each of the five treatments were independently destemmed and crushed using a crusher–destemmer (Bucher Vaslin, Niederweningen, Switzerland), and placed separately in individual 60-L plastic containers (Speidel, Swabia, Germany), where fermentation took place. Musts were inoculated with a commercial wine yeast (*Saccharomyces cerevisiae*, EC-1118, Lallemand, Rexdale, ON, Canada) at a rate of 30 g/hL. Musts were inoculated with commercial malolactic bacteria 48 h after crushing to ensure a standardized fermentation across treatments. In 2016, musts were inoculated with VP-41 (*Oenococcus oeni*, Scott Laboratories, CA, USA); in 2017, musts were inoculated with ML Prime (*Lactobacillus plantarum*, Lallemand, Rexdale, ON, Canada). Temperature and Brix were followed daily during alcoholic fermentation using a density meter (Anton Paar, Graz, Austria) with temperature and sugar consumption curves showing good reproducibility within replicates of the same and different treatments (data not shown). Following 10 days of maceration, wines were drained off from solids, with free run wines immediately transferred to 20-L glass carboys fitted with airlocks until the completion of malolactic fermentation. Following the completion of malolactic fermentation, wines were adjusted to 0.3 mg/L molecular SO_2, bottled using a DIAM 5 microagglomerated cork closure (G3 Enterprises, Modesto, CA, USA) and kept in cellar-like conditions until analysis.

2.3. Fruit Composition

Berry chemistry and physical properties were measured at harvest from random samples of 250 berries taken from each replication ($n = 3$). Brix was measured using a density meter (Anton Paar, Graz, Austria); titratable acidity (TA) was measured by titrating a known quantity of juice (5 mL) in a deionized water solution against 0.067 N NaOH (Fisher Scientific, Waltham, MA, USA) to a pH endpoint of 8.2 in accordance with an established procedure (Iland et al., 2004); pH was measured with a Benchtop pH meter (ThermoFisher Scientific, Waltham, MA, USA. Yeast assimilable nitrogen (YAN) was measured enzymatically from juice utilizing an analyzer and commercially available kits (Biosystems, Barcelona, Spain). Individual vine pruning weights were taken during vine dormancy and compared with individual vine fruit yields to calculate vine Ravaz Index [31,32].

2.4. Wine Composition

Wine titratable acidity (TA) and pH were measured in the same method as juice TA and pH. Wine ethanol was measured with an alcolyzer wine M/ME analysis system (Anton Paar, Graz, Austria); wine residual sugars, acetic acid, lactic acid, and malic acid were measured with a Y15 analyzer (Biosystems, Barcelona, Spain) using commercial enzymatic analysis kits (Biosystems, Barcelona, Spain). Wine phenolics and color were measured at pressing and following the completion of malolactic fermentation (malic acid < 0.4 g/L). Anthocyanins, non-tannin phenolics, small polymeric pigments (SPP), large polymeric pigments (LPP), and total polymeric pigments (herein reported as SPP + LPP), were measured as previously described [49]. Tannins in the wines were analyzed by protein precipitation [50]. Full-visible-spectrum absorbance scans were taken using a spectrophotometer (Cary UV-VIS60, Agilent Technologies, Santa Clara, CA, USA) and absorbance data is used to construct visible light absorbance curves and run through Cary WINUV Color module software (Agilent Technologies, Santa Clara, CA, USA) to extract CIE-L*a*b* tri-stimulus colorimetry values (D65 illuminant).

2.5. Duo-Trio Test

The 2016 wines were analyzed three months after bottling for overall sensory difference using a duo-trio test with constant reference as described [51]. Briefly, the test was administered to 21 enology students of the Wine and Viticulture Department, Cal Poly San Luis Obispo. After a brief training session (1 h) to familiarize the subjects with the test, each subject received three consecutive flights, each containing three wine ISO glasses (Libbey, Toledo, OH, USA). One glass was labeled as R ("reference") and the other two glasses were labeled with three-digit random code numbers. All the treatment replicates were contrasted against all the replicates of the control treatment. Significance was established at $p < 0.05$ and for $n = 21$, 15 correct responses were needed to establish an overall sensory difference between any of the control wines and any of the cluster thinning treatments [51]. The wines were poured 30 min before the sessions and glasses were covered with plastic lids to trap volatiles, with each glass receiving exactly 25 mL of wine.

2.6. Statistical Analyses

Statistical analyses were carried out with JMP (SAS Institute, Cary, NC, USA) for analyses of variance (ANOVA). Fisher's least significant difference (LSD) test at the $\alpha = 0.05$ level was used for means separation. Two-way ANOVA models considering treatment and growing season were carried out for all parameters with data from both growing seasons.

3. Results

3.1. Seasonal Climate

Weather data from the San Luis Obispo CIMIS weather station (Station 52, located 14.41 km from the vineyard site) was used to calculate climatological parameters during the study (Table 1). Growing degree days (GDD) were calculated for each season of the study. While GDD accumulation in 2016 placed the vineyard site in region II, considered to be a 'cool climate', there was sufficient heat in 2017 during the growing season to classify San Luis Obispo as region III, which corresponds with a 'moderately warm' climate [48].

Table 1. Growing degree days (GDD); Winkler region classification; and precipitation for San Luis Obispo, California (USA).

Year	Growing Degree Days (GDD) [1]	Winkler Region	Annual Precipitation (mm) [2]	Seasonal Precipitation (mm) [3]
2016	1462.1	II	521.9	66.2
2017	1780.6	III	733.6	79.7

[1] Calculated from 1 April–31 October in Celsius units with a baseline of 10 °C; [2] Sum of precipitation from 1 January–31 December; [3] Sum of precipitation from 1 April–31 October.

3.2. Yield

As intended, all cluster thinning treatments resulted in a reduction of clusters per vine and vine fruit yield relative to the unthinned control (Table 2). There was no significant difference in either cluster number or vine yield between thinning treatments, indicating that the thinning treatment was evenly applied. Cluster weight was lower in bloom + 8 relative to control fruit (Table 2). However, the growing season had a larger effect on cluster weight than the thinning treatments, with cluster weights being significantly higher in the warmer 2017 growing season than in the cooler 2016 growing season (Table A1 in Appendix A). Berry weight was generally unaffected by the timing of cluster thinning, with the effect of the growing season having a significant impact on this parameter. Indeed, berry weight was generally higher in the cooler 2016 growing season (Table A1). No significant treatment × growing season interaction was found in any yield component, indicating that the effect of the timing of cluster thinning on yield components was equivalent in both the cooler 2016 and warmer 2017 growing seasons. In addition, in 2017, pruning weights were collected and the Ravaz index was calculated to assess vine balance. Non-thinned control vines had a Ravaz Index of 3.23 (Table A1). The Ravaz Index was lower in bloom + 4 vines relative to control vines; bloom + 8 and bloom + 12 vines were indistinguishable from control vines or one another (Table A1).

Table 2. Two-way analysis of variance (ANOVA) with interaction showing mean values and *p*-values of vine yield components by cluster thinning treatment. Combined two-year averages followed by standard error of the mean. Also shown are *p*-values corresponding to main effects and the interaction between treatments and growing season. Different letters within a column indicate significant differences between treatment groups for Fisher's least significant difference (LSD) test at $p < 0.05$. *p* values below 0.05 are shown in bold fonts.

Treatment	Clusters per Vine	Yield per Vine (kg)	Cluster Weight (g)	Berry Weight (g)
Control	32.13 ± 3.05 a	2.43 ± 0.34 a	74.80 ± 6.35 a	1.08 ± 0.06 a
Bloom + 4	21.06 ± 3.03 b	1.27 ± 0.14 b	64.42 ± 5.98 ab	1.05 ± 0.03 a
Bloom + 8	21.52 ± 0.99 b	1.34 ± 0.15 b	58.82 ± 5.78 b	0.96 ± 0.03 a
Bloom + 12	22.10 ± 1.59 b	1.44 ± 0.15 b	65.91 ± 6.35 ab	1.07 ± 0.07 a
Treatment (T)	**0.0228**	**0.0029**	0.2179	0.3189
Growing Season (S)	0.7436	0.0569	**0.0008**	**0.0489**
T × S Interaction	0.8149	0.6569	0.8883	0.5432

3.3. Fruit Composition

All treatments were harvested manually on a single day in both 2016 and 2017. Table 3 shows Brix, titratable acidity (TA), and pH, which were determined at harvest. No significant difference was found in Brix level at harvest between treatments or between growing seasons (Table 3), indicating no effect of cluster thinning at any point during the growing season on the ability of fruit to ripen at this site. Fruit pH was lower in bloom + 12 relative to all other treatments and the non-thinned control (Table 3). There was a significant effect of the growing season on fruit pH, indicating that the growing season had a larger effect than treatments on fruit pH, with the cooler season resulting in lower fruit pH (Table 2). There was no significant treatment × growing season interaction on fruit Brix or pH, suggesting that the effect of cluster thinning timing (or lack thereof) was the same across both a warm growing season (2017) and a cool growing season (2016). No consistent effect of treatment on fruit TA was seen across both growing seasons (Table 3). In 2016, which was the cooler growing season, bloom + 12 showed higher fruit TA relative to control fruit and in 2017, the warmer growing season, bloom + 12 had lower fruit TA relative to control fruit (Table 2). However, no difference was observed in 2017 (Table 2). There was a significant interaction of treatment and growing season, indicating that the effect of cluster thinning timing on fruit TA was dependent on environmental factors pertaining to the climate of the individual growing seasons.

Table 3. Two-way analysis of variance (ANOVA) with interaction showing mean values and *p*-values of fruit composition parameters at harvest by cluster thinning treatment. Combined two-year averages followed by standard error of the mean. Also shown are *p*-values corresponding to main effects and the interaction between treatments and growing season ($n = 3$). Different letters within a column indicate significant differences between treatment groups for Fisher's LSD at $p < 0.05$. *p* values below 0.05 are shown in bold fonts.

Treatment	Brix	pH	Titratable Acidity (g/L)
Control	22.28 ± 0.30 a	3.54 ± 0.03 a	6.35 ± 0.17 a
Bloom + 4	22.42 ± 0.24 a	3.53 ± 0.04 ab	6.37 ± 0.12 a
Bloom + 8	22.72 ± 0.14 a	3.55 ± 0.04 a	6.35 ± 0.14 a
Bloom + 12	22.22 ± 0.19 a	3.48 ± 0.05 b	6.59 ± 0.50 a
Treatment (T)	0.3237	0.0896	0.725
Growing Season (S)	0.4125	**<0.0001**	**0.0115**
T × S Interaction	0.0585	0.3955	**0.001**

3.4. Wine Composition

3.4.1. Wine Basic Chemistry

Basic wine chemistry was analyzed on the finished wines at the time of bottling. Bloom + 4 wine had higher pH relative to the control and bloom + 12 wines (Table 4). The growing season also significantly affected wine pH (Table 4), with pH being higher in the warmer growing 2017 season. However, there was not a significant treatment × growing season interaction, indicating that while seasonal variation in environment did influence wine pH, the effect of cluster thinning on wine pH was not affected by seasonal variation. No effect of thinning treatment was found on wine TA (Table 4). Wine TA levels were significantly higher in 2016 than 2017, indicating an effect of the growing season (Table 4). No effect of thinning treatment was found on ethanol (Table 4). However, wine ethanol levels were higher in 2016 than in 2017 (Table 4), indicating, once again, a significant effect of the growing season over the cluster thinning treatments on the basic chemistry of the resulting wines. Bloom + 4 and bloom + 8 wines had significantly higher acetic acid levels relative to control and bloom + 12 wines (Table 4), which were statistically indistinguishable from one another. Bloom + 4 and bloom + 8 thinning treatments resulted in higher wine acetic acid levels relative to bloom + 12 and control wines (Table 4), which were statistically indistinguishable from one another. Wine acetic acid was lower in 2017 relative to 2016 (Table 4), and a significant interaction of treatment and growing season was observed (Table 4), indicating that seasonal variation in climate affected the impact of cluster thinning treatments on wine acetic acid. Although there were differences observed in wine acetic acid, because of the low nature of the acetic acid levels in the wines, these differences are unlikely to be of sensory relevance.

3.4.2. Wine Phenolics

Wine phenolics, including anthocyanins, tannins, polymeric pigments, and total phenolics, were measured as previously described [49]. There was no effect of cluster thinning treatment on wine anthocyanins or polymeric pigments in either growing season (Table 5). Total phenolics, the summation of total tannins and non-tannin phenolics, was significantly higher in bloom + 4 relative to other cluster thinning treatments, although none of the treatments were statistically distinguishable from the control (Table 5). Polymeric pigments, tannins, and non-tannin phenolics exhibited greater differences due to seasonal variation than to the cluster thinning treatment (Table 5), with polymeric pigments and tannins being lower in the warmer season than in the cooler season and non-tannin phenolics being higher in the warmer season (Table 6).

3.4.3. Wine Color

Wine CIE L*a*b* color space values were determined in the finished wines at bottling. Bloom + 12 wines exhibited higher a* and chroma than bloom + 4 and bloom + 8 wines, but all were statistically indistinguishable from control wines (Table 6). No effect of cluster thinning treatment was found in wine L*, b*, or hue angle. The growing season had a comparatively higher impact than cluster thinning treatment in every chromatic parameter (Table 6), with 2016 wines having lower L*, b*, and hue angle than 2017 wines, and higher a* and chroma than 2017 wines (Table 5). No treatment × growing season interaction was found in any chromatic parameter.

Table 4. Two-way analysis of variance (ANOVA) with interaction showing mean values and *p*-values of wine composition parameters post-malolactic fermentation by cluster thinning treatment. Combined two-year averages followed by standard error of the mean. Also shown are *p*-values corresponding to main effects and the interaction between treatments and growing season (*n* = 3). Different letters within a column indicate significant differences between treatment groups for Fisher's LSD at *p* < 0.05. *p* values below 0.05 are shown in bold fonts.

Treatment	pH	Ethanol (% v/v)	Titratable Acidity (g/L)	Acetic Acid (g/L)	Lactic Acid (g/L)	Malic Acid (g/L)	Residual Sugar (g/L)
Control	3.81 ± 0.03 b	13.03 ± 0.16 a	5.36 ± 0.29 a	0.25 ± 0.04 b	1.23 ± 0.06 b	0.09 ± 0.03 a	0.45 ± 0.03 ab
Bloom + 4	3.86 ± 0.03 a	13.05 ± 0.16 a	4.91 ± 0.25 a	0.31 ± 0.03 a	1.32 ± 0.06 a	0.09 ± 0.02 a	0.43 ± 0.02 ab
Bloom + 8	3.84 ± 0.03 ab	13.17 ± 0.16 a	5.21 ± 0.24 a	0.31 ± 0.02 a	1.25 ± 0.07 b	0.04 ± 0.02 a	0.41 ± 0.02 b
Bloom + 12	3.77 ± 0.02 c	13.24 ± 0.12 a	5.25 ± 0.16 a	0.24 ± 0.04 b	1.32 ± 0.05 a	0.06 ± 0.03 a	0.47 ± 0.01 a
Treatment (T)	**0.0002**	0.4711	0.2637	**<0.0001**	**0.0046**	0.4302	0.1378
Growing Season (S)	**<0.0001**	**0.0002**	**<0.0001**	**<0.0001**	**<0.0001**	0.6511	**0.0024**
T × S Interaction	0.2238	0.7141	0.6896	**0.0031**	0.2296	0.5036	0.8459

Table 5. Two-way analysis of variance (ANOVA) with interaction showing mean values and *p*-values of wine phenolic parameters post-malolactic fermentation by cluster thinning treatment. Combined two-year averages followed by standard error of the mean. Also shown are *p*-values corresponding to main effects and the interaction between treatments and growing season (*n* = 3). Different letters within a column indicate significant differences between treatment groups for Fisher's LSD at *p* < 0.05. *p* values below 0.05 are shown in bold fonts.

Treatment	Anthocyanins (mg/L Malvidin Equivalents)	Polymeric Pigments (Absorbance at 520 nm)	Tannins (mg/L CE[1])	Non-Tannin Phenolics (mg/L CE[1])	Total Phenolics (mg/L CE[1])
Control	193.29 ± 2.80 a	0.86 ± 0.14 a	22.85 ± 7.43 ab	520.11 ± 20.64 ab	542.96 ± 22.86 ab
Bloom + 4	197.43 ± 4.00 a	0.95 ± 0.11 a	26.99 ± 4.35 a	553.86 ± 11.40 a	580.85 ± 8.85 a
Bloom + 8	197.56 ± 3.96 a	0.82 ± 0.10 a	16.64 ± 1.42 ab	516.66 ± 16.87 b	533.30 ± 16.29 b
Bloom + 12	197.22 ± 5.43 a	0.85 ± 0.10 a	14.66 ± 0.94 b	522.81 ± 10.91 ab	537.46 ± 10.92 b
Treatment (T)	0.8413	0.4305	0.1078	0.1629	0.1175
Growing Season (S)	0.2201	**<0.0001**	**0.0304**	**0.0183**	0.1208
T × S Interaction	0.2255	0.8451	0.1186	0.0672	0.2156

[1] CE: Catechin equivalents.

Table 6. Two-way analysis of variance (ANOVA) with interaction showing mean values and *p*-values of wine CIE L*a*b* chromatic parameters post-malolactic fermentation by cluster thinning treatment. Combined two-year averages followed by standard error of the mean. Also shown are *p*-values corresponding to main effects and the interaction between treatments and growing season (*n* = 3). Different letters within a column indicate significant differences between treatment groups for Fisher's LSD at *p* < 0.05. *p* values below 0.05 are shown in bold fonts.

Treatment	L*	a*	b*	Hue Angle	Chroma
Control	85.34 ± 2.14 a	18.37 ± 2.60 ab	2.51 ± 1.53 a	−0.86 ± 0.60 a	19.03 ± 2.31 ab
Bloom + 4	85.67 ± 2.32 a	17.67 ± 2.79 b	2.55 ± 1.38 a	−0.34 ± 0.54 a	18.31 ± 2.51 b
Bloom + 8	85.85 ± 2.24 a	17.71 ± 2.61 b	2.65 ± 1.39 a	−0.28 ± 0.49 a	18.36 ± 2.32 b
Bloom + 12	84.37 ± 1.80 a	19.57 ± 2.46 a	2.71 ± 1.62 a	−0.77 ± 0.54 a	20.24 ± 2.17 a
Treatment (T)	0.1982	0.1149	0.8018	0.4819	0.1072
Growing Season (S)	**<0.0001**	**<0.0001**	**<0.0001**	**<0.0001**	**<0.0001**
T × S Interaction	0.3599	0.8472	0.067	0.5101	0.8306

3.5. Duo-Trio Test

Figure 1 shows a duo-trio test performed in the 2016 wines three months after bottling. Under the "constant reference" variant of this test, each wine treatment is contrasted against the control [51]. The results of this test indicated that none of the cluster thinning treatments produced wines that had overall sensory differences relative to the control wine. The panel ($n = 22$) failed to find an overall sensory difference between the control and any of the cluster thinning wines.

Based on these results, and considering that a panel of four experience industry professionals deemed the differences in the 2017 wines to be even less evident than those in 2016, no sensory analysis was performed in the 2017 wines.

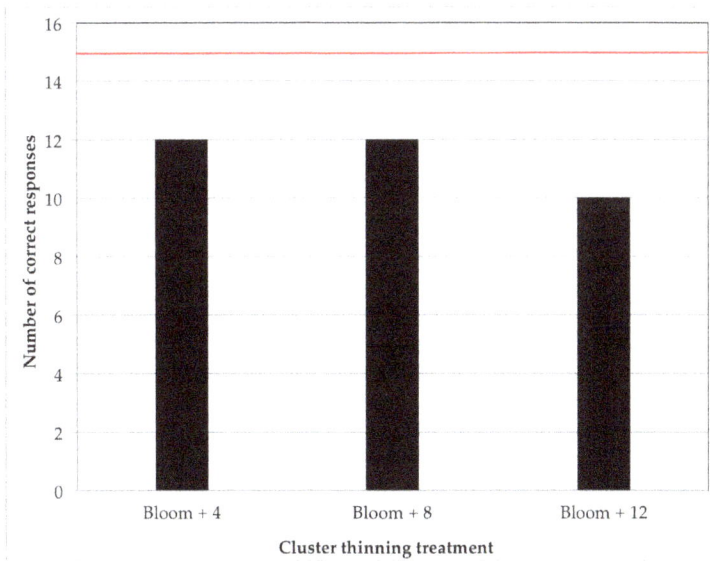

Figure 1. Results of the duo-trio test with constant reference performed in the 2016 wines. Each cluster thinning treatment was contrasted against the control treatment. A total of 15 correct responses were needed for a statistically distinguishable sensory difference ($p < 0.05$). The horizontal red line indicates the number of correct responses required to attain statistical significance.

4. Discussion

A study was conducted over two growing seasons (2016 and 2017) to determine the agronomical effects (on grapes) and chemical effects (on the resulting wines) of cluster thinning timing in Pinot noir grown in the Edna Valley of California's central coast. A secondary objective of the study was to identify appropriate crop loads for Pinot noir on the moderately cool climate of California's central coast. Cluster thinning treatments were applied at 4, 8, and 12 weeks post-bloom, approximating the timing of the phenological growth events of fruit set and véraison and including a pre-harvest "red drop". Thinning consisted of removing any second or third cluster from fruit bearing shoots, reducing cluster number by an average of 34.3% across all treatments (Table 2), and reducing yield by an average of 44.4% across all treatments (Table 2). Previous studies conducted on cluster thinning practices have applied variable yield reduction rates depending on cultivar. However, most research in Pinot noir has applied "half crop" treatments such as the one performed in the current study, removing all but the basal cluster on each fruiting shoot [34] or reducing cluster number by 50% [19].

Growing degree day (GDD) accumulation varied between 2016 and 2017 enough to place each growing season in a separate Winkler Index region, with 2017 being warmer by 318.5 GDD than 2016 and qualifying as a region III (Table 1). An increase in average temperature by 1 °C increases GDD by 214 over the course of a growing season. The GDD variation observed at this site between 2016 and 2017 corresponds to approximately 1.5 °C higher average temperatures in 2017. It is well documented that vine capacity and the ability to achieve ripeness by Brix accumulation in cool regions is dependent upon climatic conditions such as temperature [48,52,53], with a base level of temperature required to adequately ripen fruit. However, Brix accumulation is not driven solely by temperature and, by extension, by GDD accumulation. In fact, differences in berry temperature may not directly impact Brix accumulation at all [52]. Indeed, other factors such as soil moisture content [54] and berry light exposure [55] may also impact fruit ripening rate irrespective of atmospheric temperature and may even have a larger effect than GDD accumulation in cases where GDD accumulation is adequate for fruit ripening [54]. Consequently, in the present study, there was no significant effect of the growing season on Brix at harvest to indicate an impact of the increased GDD accumulation in 2017 (Table 3), despite the fruit being picked on the same date in both growing seasons.

In the present study, there was no impact of the thinning treatments on Brix accumulation. However, the climatic conditions of the growing season did have a clear impact on fruit pH and TA (Table 3), resulting in lower TA and higher pH in 2017 fruit (Table 2). While bloom + 12 fruit did have significantly lower pH than control fruit (Table 3), after accounting for growing season in the model, the effect of cluster thinning treatment was not significant, indicating that the climatic conditions prevalent during the growing season had a greater effect on fruit pH and TA than cluster thinning timing. The decrease of malate concentration in grapes post-véraison through respiration increases with temperature and light exposure [56,57]; as such, the decrease in TA and the corresponding increase in pH observed in 2017, which was the warmer season (Table 2), is likely a function of the increased temperature. There was a significant treatment × growing season interaction found in TA (Table 3). In both growing seasons, bloom + 12 fruit had significantly different TA from the control treatment. However, the direction of the difference varied, with bloom + 12 having higher TA in 2016 and lower TA in 2017 relative to the control treatment. In both growing seasons, bloom + 4 was indistinguishable from the control treatment. Bloom + 8 was indistinguishable from the control or any treatment in 2016 and significantly lower than the control in 2017. Overall, there was no consistent effect of cluster thinning on fruit TA across both growing seasons (Table 3). Previous studies on Pinot noir, in which half of the crop was thinned, have found increased pH [19,53] in both fruit and juice, but mixed effects on TA, showing either no impact [19] or a reduction in TA with cluster thinning [53]. While no significant treatment effect of pH was found in the present study after considering the growing season, the inconsistent results observed in bloom + 12 fruit is similar to what has been found in previous research, suggesting the influence of some external factor on malate degradation in late-thinned fruit. It is possible that late thinning resulted in more convective heat exchange between clusters and air within the canopy, which, given the substantially warmer air temperatures in August of 2017 (Table 7), resulted in an increase in berry temperature and therefore malate degradation in the 2017 bloom + 12 fruit that was not seen in 2016. Indeed, cooler air temperatures in 2016 during the same period (Table 7) may have resulted in decreased berry temperatures and the observed increase in fruit TA relative to the non-thinned control.

Wine pH is tied intrinsically to a wine's microbial and oxidative stability, with lower pH values inhibiting (synergistically with ethanol) microbial growth [58] and increasing the ability of phenolics to protect the wine from premature oxidation [59]. There was a significant effect of thinning timing on wine pH, with bloom + 12 having lower wine pH relative to the control in accordance with the observed lowered fruit pH. Conversely, the bloom + 4 wines showed higher wine pH relative to the control wines despite indistinguishable fruit pH (Tables 3 and 4). The growing season also significantly affected wine pH (Table 4) with the warmer 2017 growing season resulting in higher wine pH than 2016, much like the effect seen on fruit pH (Table 2). However, the interaction for treatment and growing season was not significant, suggesting that while seasonal variances in environmental factors do influence wine pH, environmental variance did not affect the response of wine pH to thinning timing. Interestingly, differences in fruit pH did not correspond linearly with differences in wine pH. Both growing season and treatment affected wine lactic acid content; fruit malic acid was measured in 2017 and no significant difference was found between treatments ($p = 0.38$, df = 4.7; data not shown), indicating that a difference in fruit malic acid content was not responsible for the differences observed in lactic acid content in 2017. Different strains of malolactic bacteria were used in 2016 and 2017; VP-41 (*Oenococcus oeni*) in 2016 and ML-Prime (*Lactobacillus Plantarum*) in 2017. Unrelated to malic acid content, *Lactobacillus* and *Oenococcus* fermentation activity in wine is affected by wine temperature, ethanol level, pH, and acetic acid levels [60], each of which exhibited some degree of variation within the wines that could be responsible for the differences observed between treatment groups in wine lactic acid and pH levels, irrespective of differences in fruit composition. Lower average and maximum fermentation temperatures in 2017 wines corresponded with higher wine pH and higher acetic acid levels (Tables 3 and 4), indicating that fermentation temperature was a likely contributor to differences in wine pH and acetic acid levels. Indeed, a two-way ANOVA utilizing treatment, average temperature, maximum temperature, and treatment × average temperature and treatment × maximum temperature interactions, found average temperature to be a significant predictor of wine pH ($p = 0.0136$) and acetic acid ($p = 0.0136$).

Despite no differences being found in fruit Brix (Table 3), wine ethanol was significantly higher in 2016 than 2017 (Table 4). As yeast and fermentation practices were constant between growing seasons, observed differences in ethanol content are most likely the result of variations of the alcohol conversion ratio of the yeast. Average must temperature during the 10-day maceration period in 2016 was 21.3 °C and 24.9 °C in 2017, with peak fermentation temperatures of 26.7 °C and 29.9 °C, respectively (Table 3). Similar to *Lactobacillus* and *Oenococcus* fermentation activity, temperature is one of the most influential factors on *Saccharomyces cerevisiae* fermentation activity and ethanol biosynthesis [61,62], with ethanol formation decreasing as fermentation temperature increases [61]. In addition to decreasing ethanol biosynthesis in wine yeast, increased fermentation temperatures also increase the rate of ethanol volatilization, further lowering already diminished wine ethanol levels [61].

The growing season also affected cluster and berry weight (Table 2), with 2017 resulting in fruit having 34% higher cluster weight and 9% lower berry weight relative to 2016 (Table A1). Several factors influence berry size, including berry temperature during various phenological growth stages, light incidence, water and nutrient supply, and seed number per berry [63–66]. Berry size can be reduced by increased heat and resultant increased berry temperature prior to the lag phase [64]. Ambient temperature was on average 1.25 °C warmer in May 2017 than May 2016 (Table 7), which may explain current results. Cluster weight is a function of berry size and berry number, so as berry weight decreased while cluster weight increased, berry number must have increased in 2017. Typical berry set in wine grapes ranges from 20% to 50% [67], and can be reduced by temperatures during bloom below 15 °C [64]. From 15 April to 1 June, there were 29 days with an average air temperature below 15 °C in 2016 and 14 days in 2017 (Table 7). It is likely that warmer temperatures during bloom in 2017 resulted in a higher fruit set and therefore higher cluster weight than 2016.

Pruning weights were collected in January 2018 and Ravaz Index was calculated for the 2017 growing season. The non-thinned control vines had a Ravaz Index of 3.23, and Ravaz Index was not significantly different between treatments (Table A1). Within the control vines, one replication had a substantially lower cluster number than other replications, which inflated the deviation of cluster number, vine yield, and Ravaz Index of the sampled population (Table A1). This was confirmed by conducting outlier analysis of control treatment repetition 1, which, for the Ravaz Index model, had a Cook's Distance value of 16 (data not shown), indicating high influence on the model. It is possible that the low cluster number on the vines within this repetition is due to natural site variation (e.g., block to block variations in soil composition or water holding capacity), affecting vine capacity. While the abnormally low cluster number affected vine yield, cluster number, and Ravaz Index, little influence of this repetition was found in models of fruit or wine chemistry, with no Cook's distance value above 0.5 (data not shown). As a result of the potential impact of eliminating one of the three replications of the control treatment from the dataset on experimental balance and statistical analysis, the outlier was retained within the dataset.

Phenolic composition (Table 5) and wine color (Table 6) were not affected consistently by any of the thinning treatments, but all chromatic parameters as well as wine polymeric pigment, tannin, and non-tannin phenolic content were significantly affected by growing season. Polymeric pigments, formed by the covalent polymerization of anthocyanins with monomeric flavan-3-ols or tannins [8] provide protection for anthocyanins against oxidation [68,69], which can be beneficial in wines made from cultivars lacking in phenolics, such as Pinot noir. However, as their formation may also lower wine saturation on accounts of their lower molar extinction coefficient relative to that of intact anthocyanins [70], increasing polymeric pigments may result in comparative decreases in wine color saturation. Notwithstanding, polymeric pigments generally provide desirable mouthfeel properties as they are less astringent than tannins of the same molecular weight [8]. Polymeric pigments and tannins were lower in 2017, while non-tannin phenolics were higher (Table 6). While some parameters (total phenolics, non-tannin phenolics, L*, a*, chroma) were affected by thinning treatment in 2017 (Table 6), no consistent effect of treatment or treatment × growing season interaction was found in any wine phenolic or chromatic parameter. Polymeric pigments were likely higher in 2016 because of the increased level of tannins observed (Table 6), despite no difference in anthocyanin levels between growing seasons.

Wine CIE L*a*b* color parameters L*, b*, and hue angle were higher in 2017 than 2016, while a* and chroma were lower, indicating wine color was darker and bluer in the wines from the warmer growing season, but less saturated and red than wines from the cooler growing season. The color shift observed in 2017 wines is likely due to differences in wine pH and polymeric pigment content. The effect of pH on wine color and anthocyanin chromatic parameters is well established. As pH decreases, the equilibrium of anthocyanin forms shifts to favor the flavilium cation, which increases red hue, and as pH increases, the equilibrium shifts to favor the quinonoidal hydrobase, which increased blue hue [71]. Additionally, as pH increases, there is a linear decrease in chroma value observed [71]. As wine pH was generally higher in the 2017 wines (Table 4), it would follow that the color in 2017 wines, while not having statistically distinguishable anthocyanin concentration relative to 2016 wines, would have increased blue hue and lower chroma. Polymeric pigment formation may lower saturation (as indicated by chroma). Saturation may be lowered through the transformation (and subsequent reduction) of anthocyanins, or through the modulation of the chromatic properties of the anthocyanin subunit following a reduction of the molar extinction coefficient relative to the native anthocyanin, although there is only indirect experimental evidence of this molar extinction coefficient reduction [72,73].

The results of the overall sensory test performed in the 2016 wines generally mirrored previously uncovered trends in the basic, phenolic, and chromatic composition of the resulting wines. That is, none of the cluster thinning treatments produced wines that were distinguishable, from a sensory standpoint, from the control wines. Similar to what has been found in the present study, cluster thinning performed in Chardonnay Musqué grapes, while producing chemical differences in fruit, resulted in little sensory differences in the resulting wines [27]. In another study, wine produced from cluster thinned Cabernet Sauvignon vines exhibited a small increase in perceived wine quality relative to wine produced from non-thinned vines [40], although location was found to have a greater impact on sensory perception than the cluster thinning treatment, and the effect was not consistent across growing seasons. While wine chemical composition and perceived sensory attributes rarely follow linear correlations, without corresponding differences in chemical composition, it is unlikely that cluster thinning will have an impact on wine sensory perception. Therefore, the lack of sensory differences observed in the wines of the present study is unsurprising, and we hypothesize that any chemical differences in volatiles that may have occurred within the wines were below sensory thresholds, and therefore practically irrelevant. Unless cluster thinning is necessitated by the vine balance (i.e., the vine is overcropped) in cool climate Pinot noir, it is unlikely that there will be any sensory benefit to cluster thinning that would justify the negative economic impact associated with cluster thinning.

In much of the previous research performed on cluster thinning, external factors independent of (although at times in combination with) vine crop load impacted berry and wine chemical composition to a greater degree than crop load. Factors such as climatic variation in growing season [21] and viticultural practices such as floor management [34], deficit irrigation [21,22], and leaf thinning [74,75] have all been found to have a greater effect on fruit composition than cluster thinning. Indeed, in the present study, no consistent effect of cluster thinning or cluster thinning timing was observed across two growing seasons, a cooler growing season and a warmer growing season. Conversely, the growing season had a greater effect on variation in fruit and wine composition than thinning treatment for most parameters. A Ravaz Index range of 3 to 6 has been previously proposed as an optimum crop load for Pinot noir grown in cool-climates [30]. Based on the lack of differences observed in fruit Brix accumulation, wine composition, and wine sensory perception, Pinot noir vineyards on the central coast of California can, barring climatic conditions severely increasing crop set or severely limiting ripening potential, likely support higher crop levels than those of the vineyard utilized in this this study, which had Ravaz Index values of 3.23 across the non-thinned blocks. Considering previously proposed ranges and the results of the current study, a Ravaz Index value of 6 could be appropriate for Pinot noir on the central coast of California, and should be examined and evaluated accordingly in future work.

5. Conclusions

No positive effect of cluster thinning or the timing of it was observed across two growing seasons, a cooler growing season and a warmer growing season, for Pinot noir grapes and wines. In general, the growing season had a greater effect on variation than thinning treatment for most parameters. Few treatment × growing season interactions were found in wine composition parameters, indicating that rather than cluster thinning treatment being affected by seasonal variation, which has been reported previously, seasonal variation itself was the primary driver of differences in fruit and wine composition. No sensory differences were detected between the non-thinned control and any wines from cluster thinned treatments. However, on average, cluster thinning was associated with a 44% reduction in crop yields, and this reduction in crop load failed to produce a positive or discernible sensory effect on the resulting wines. Pinot noir vineyards on the central coast of California can support crop loads that result in Ravaz Index values larger than 3.23 and potentially up to 6 without concern for impacting ripening potential, barring a severe decrease in GDD accumulation. This study also suggests that in Pinot noir, balanced canopies with LA/Y ratios in tune with the prevalent seasonal conditions of the region would most likely yield quality fruit, with no discernible or marginal improvements in quality due to cluster thinning.

Author Contributions: Conceptualization, J.C.D.P. and L.F.C.; Methodology, J.C.D.P. and L.F.C.; Validation, P.F.W.M., J.C.D.P., and L.F.C.; Formal Analysis, P.F.W.M. and L.F.C.; Investigation, P.F.W.M., J.C.D.P., and L.F.C.; Writing—Original Draft Preparation, P.F.W.M.; Writing—Review & Editing, P.F.W.M., J.C.D.P., and L.F.C.; Visualization, P.F.W.M., J.C.D.P., and L.F.C.; Supervision, J.C.D.P. and L.F.C.; Project Administration, J.C.D.P. and L.F.C.; Funding Acquisition, J.C.D.P. and L.F.C.

funding: This research was funded by the California State University Agricultural Research Institute grant 2017-58874, the California Polytechnic State University, San Luis Obispo, Baker Koob Endowment grant 2018-65141, California Polytechnic State University, San Luis Obispo, College of Agriculture and Environmental Sciences Summer Undergraduate Research Program (SURP) and the Crimson Wine Group (Napa Valley, CA, USA).

Acknowledgments: The authors would like to extend their sincerest gratitude to Jordan Stanley, Claire Villaseñor, and Vegas Riffle for their assistance with this project. The authors also thank Chamisal Vineyards, especially Fintan Du Fresne, for generously offering space, time, and logistical resources to make this project possible.

Conflicts of Interest: The authors declare no conflict of interest. The founding sponsors had no role in the design of the study; in the collection, analyses, or interpretation of data; in the writing of the manuscript; and in the decision to publish the results.

Appendix A

Appendix A.1. Supplementary Fruit and Wine Data

Table A1. Vine fruit yield and fruit physical composition by treatment and growing season. Treatment means followed by standard error of the mean. Different letters within a column and growing season indicate significant differences between treatment groups for Fisher's least significant difference (LSD) test at $p < 0.05$.

Growing Season	Treatment	Clusters per Vine	Vine Yield (kg)	Cluster Weight (g)	Berry Weight (g)	Seed Weight (g)	Seeds per Berry	Pruning Weight (kg)	Ravaz Index
2016	Control	33 ± 4.93 a	2.10 ± 0.34 a	63.18 ± 1.31 a	1.18 ± 0.01 a	ND[1]	ND	ND	ND
	Bloom + 4	22.4 ± 5.29 ab	1.27 ± 0.25 b	58.86 ± 7.21 a	1.08 ± 0.04 ab	ND	ND	ND	ND
	Bloom + 8	20.1 ± 0.77 b	1.04 ± 0.16 b	51.35 ± 6.33 a	0.97 ± 0.01 b	ND	ND	ND	ND
	Bloom + 12	23.1 ± 2.74 ab	1.24 ± 0.03 b	54.70 ± 5.87 a	1.12 ± 0.10 ab	ND	ND	ND	ND
	p Values	0.1701	**0.0461**	0.519	0.0986				
2017	Control	32.30 ± 4.64 a	2.76 ± 0.59 a	86.42 ± 8.04 a	0.98 ± 0.07 a	0.060 ± 0.007 a	1.37 ± 0.05 a	0.86 ± 0.06 a	3.23 ± 0.79 a
	Bloom + 4	19.10 ± 0.95 b	1.27 ± 0.19 b	72.77 ± 9.23 a	1.01 ± 0.06 a	0.042 ± 0.009 a	1.19 ± 0.12 a	0.73 ± 0.08 a	1.67 ± 0.17 b
	Bloom + 8	23.70 ± 0.30 ab	1.63 ± 0.05 b	70.02 ± 2.07 a	0.95 ± 0.06 a	0.052 ± 0.000 a	1.44 ± 0.06 a	0.91 ± 0.06 a	1.91 ± 0.15 ab
	Bloom + 12	21.10 ± 2.03 b	1.64 ± 0.26 b	77.12 ± 6.45 a	1.01 ± 0.11 a	0.051 ± 0.004 a	1.30 ± 0.11 a	0.83 ± 0.06 a	2.01 ± 0.34 ab
	p Values	0.1179	0.0616	0.4665	0.9446	0.2773	0.3983	0.1921	0.1352

[1] ND: Not determined.

Table 2. Fruit chemical composition by treatment and growing season. Treatment means followed by standard error of the mean. Different letters within a column and growing season indicate significant differences between treatment groups for Fisher's LSD at $p < 0.05$. p values below 0.05 are shown in bold fonts. TA—titratable acidity.

Growing Season	Treatment	Fruit Brix	Fruit pH	Fruit TA (g/L)
	Control	22.73 ± 0.43 a	3.48 ± 0.04 a	6.03 ± 0.17 b
	Bloom + 4	22.67 ± 0.44 a	3.44 ± 0.03 a	6.47 ± 0.19 b
2016	Bloom + 8	22.70 ± 0.26 a	3.47 ± 0.02 a	6.63 ± 0.15 ab
	Bloom + 12	21.87 ± 0.15 a	3.37 ± 0.05 a	7.53 ± 0.53 a
	p	0.2908	0.2142	**0.0433**
	Control	21.83 ± 0.27 b	3.60 ± 0.01 a	6.68 ± 0.13 a
	Bloom + 4	22.17 ± 0.17 ab	3.61 ± 0.01 a	6.28 ± 0.15 ab
2017	Bloom + 8	22.73 ± 0.15 a	3.62 ± 0.01 a	6.08 ± 0.04 bc
	Bloom + 12	22.57 ± 0.19 a	3.59 ± 0.02 a	5.65 ± 0.28 c
	p	**0.0475**	0.2944	**0.0184**

Table 3. Fermentation Temperature by treatment and growing season. Treatment means followed by standard error of the mean. Different letters within a column and growing season indicate significant differences between treatment groups for Fisher's LSD at $p < 0.05$. p values below 0.05 are shown in bold fonts.

Growing Season	Treatment	Average Temperature (°C)	Maximum Temperature (°C)
	Control	21.10 ± 0.06 c	26.23 ± 0.15 b
	Bloom + 4	21.52 ± 0.08 a	26.57 ± 0.03 b
2016	Bloom + 8	21.44 ± 0.12 ab	27.07 ± 0.07 a
	Bloom + 12	21.18 ± 0.12 bc	26.20 ± 0.25 b
	p	**0.0456**	**0.0117**
	Control	25.08 ± 0.12 a	31.17 ± 0.50 a
	Bloom + 4	24.69 ± 0.10 b	29.20 ± 0.35 b
2017	Bloom + 8	24.58 ± 0.04 b	28.97 ± 0.23 b
	Bloom + 12	25.28 ± 0.06 a	30.73 ± 0.27 a
	p	**0.0014**	**0.0049**

Table 4. Wine chemical composition post-malolactic fermentation by treatment and growing season. Treatment means followed by standard error of the mean. Different letters within a column and growing season indicate significant differences between treatment groups for Fisher's LSD at $p < 0.05$. p values below 0.05 are shown in bold fonts.

Growing Season	Treatment	L-Malic (g/L)	L-Lactic (g/L)	Residual Sugar (g/L)	Acetic Acid (g/L)	EtOH (% v/v)	pH	Titratable Acidity (g/L)
2016	Control	0.08 ± 0.00 a	1.12 ± 0.03 bc	0.48 ± 0.04 a	0.34 ± 0.01 ab	13.18 ± 0.32 a	3.75 ± 0.01 a	5.67 ± 0.03 a
	Bloom + 4	0.07 ± 0.04 a	1.18 ± 0.00 ab	0.46 ± 0.02 a	0.35 ± 0.01 a	13.35 ± 0.14 a	3.79 ± 0.02 a	5.47 ± 0.07 b
	Bloom + 8	0.07 ± 0.02 a	1.11 ± 0.00 c	0.45 ± 0.03 a	0.35 ± 0.01 a	13.48 ± 0.16 a	3.77 ± 0.02 a	5.67 ± 0.03 a
	Bloom + 12	0.04 ± 0.01 a	1.23 ± 0.03 a	0.49 ± 0.02 a	0.32 ± 0.01 b	13.50 ± 0.02 a	3.73 ± 0.02 a	5.60 ± 0.06 ab
	p	0.6627	**0.0111**	0.8018	0.062	0.6209	0.214	0.0672
2017	Control	0.11 ± 0.06 a	1.35 ± 0.05 ab	0.42 ± 0.04 ab	0.16 ± 0.01 b	12.88 ± 0.10 a	3.86 ± 0.01 b	5.05 ± 0.56 a
	Bloom + 4	0.10 ± 0.04 a	1.46 ± 0.01 a	0.39 ± 0.01 ab	0.26 ± 0.03 a	12.74 ± 0.14 a	3.94 ± 0.01 a	4.35 ± 0.09 a
	Bloom + 8	0.01 ± 0.01 a	1.40 ± 0.02 a	0.37 ± 0.01 b	0.28 ± 0.01 a	12.87 ± 0.07 a	3.91 ± 0.01 a	4.75 ± 0.26 a
	Bloom + 12	0.08 ± 0.06 a	1.42 ± 0.03 b	0.45 ± 0.01 a	0.16 ± 0.01 b	12.98 ± 0.09 a	3.81 ± 0.02 c	4.90 ± 0.05 a
	p	0.4522	0.1326	0.1003	**0.0021**	0.5106	**0.0006**	0.4772

Table 5. Wine chromatic parameters post-malolactic fermentation by treatment and growing season. Treatment means followed by standard error of the mean. Different letters within a column and growing season indicate significant differences between treatment groups for Fisher's LSD at $p < 0.05$. p values below 0.05 are shown in bold fonts.

Growing Season	Treatment	L*	a*	b*	Hue	Chroma
2016	Control	80.63 ± 0.75 a	24.13 ± 0.69 a	−0.90 ± 0.12 a	−2.17 ± 0.32 a	24.13 ± 0.69 a
	Bloom + 4	80.60 ± 1.01 a	23.73 ± 1.39 a	−0.50 ± 0.40 a	−1.13 ± 0.92 a	23.73 ± 1.39 a
	Bloom + 8	80.93 ± 0.86 a	23.47 ± 0.90 a	−0.43 ± 0.35 a	−1.00 ± 0.81 a	23.47 ± 0.90 a
	Bloom + 12	80.50 ± 1.03 a	24.90 ± 1.26 a	−0.90 ± 0.12 a	−1.97 ± 0.18 a	24.90 ± 1.26 a
	p	0.988	0.8069	0.5211	0.5131	0.8069
2017	Control	90.05 ± 0.24 a	12.61 ± 0.43 b	5.91 ± 0.19 ab	0.44 ± 0.02 a	13.94 ± 0.40 b
	Bloom + 4	90.74 ± 0.54 a	11.61 ± 0.36 b	5.60 ± 0.11 b	0.45 ± 0.00 a	12.89 ± 0.37 b
	Bloom + 8	90.76 ± 0.32 a	11.96 ± 0.39 b	5.72 ± 0.14 b	0.45 ± 0.01 a	13.26 ± 0.41 b
	Bloom + 12	88.24 ± 0.38 b	14.25 ± 0.45 a	6.31 ± 0.15 a	0.42 ± 0.01 a	15.59 ± 0.42 a
	p	**0.0053**	**0.0079**	**0.0402**	0.2863	**0.0062**

Table 6. Wine Phenolic profile post-malolactic fermentation by treatment and growing season. Treatment means followed by standard error of the mean. Different letters within a column and growing season indicate significant differences between treatment groups for Fisher's LSD at $p < 0.05$. p values below 0.05 are shown in bold font.

Growing Season	Treatment	Total Anthocyanins (mg/L Malvidin Equivalents)	Total Polymeric Pigments (Absorbance at 520 nm)	Total Tannins (mg/L Catechin Equivalents)	Total Phenolics (mg/L Catechin Equivalents)	Non-Tannin Phenolics (mg/L Catechin Equivalents)
2016	Control	191.41 ± 4.09 a	1.13 ± 0.15 a	31.70 ± 14.01 a	519.61 ± 45.10 a	487.91 ± 32.40 a
	Bloom + 4	199.07 ± 8.70 a	1.16 ± 0.05 a	36.35 ± 2.36 a	574.75 ± 16.31 a	538.40 ± 17.35 a
	Bloom + 8	197.44 ± 8.52 a	1.04 ± 0.07 a	17.36 ± 2.86 a	503.71 ± 18.50 a	486.35 ± 20.19 a
	Bloom + 12	187.57 ± 7.13 a	1.07 ± 0.06 a	13.19 ± 1.34 a	549.21 ± 16.60 a	536.03 ± 15.42 a
	p	0.6755	0.7662	0.1500	0.3215	0.2481
2017	Control	195.17 ± 4.37 b	0.59 ± 0.01 a	14.00 ± 1.08 b	566.31 ± 5.92 a	552.31 ± 6.52 a
	Bloom + 4	195.78 ± 1.22 b	0.74 ± 0.13 a	17.63 ± 1.22 a	586.96 ± 9.42 a	569.32 ± 10.49 a
	Bloom + 8	197.67 ± 2.37 b	0.61 ± 0.01 a	15.92 ± 1.15 ab	562.89 ± 10.46 a	546.97 ± 9.83 a
	Bloom + 12	206.86 ± 1.85 a	0.64 ± 0.02 a	16.13 ± 0.67 ab	525.71 ± 13.49 b	509.59 ± 13.52 b
	p	0.0522	0.4078	0.1905	**0.0171**	**0.0203**

A.2. Weather Data

Table 7. Daily weather data from California Irrigation Information Management System (CIMIS) weather station 52 in San Luis Obispo during the 2016 and 2017 growing seasons. Daily average air temperature, minimum air temperature, and maximum air temperature.

Date	2016 Average Air Temperature (°C)	2016 Minimum Air Temperature (°C)	2016 Maximum Air Temperature (°C)	2017 Daily Average Air Temperature (°C)	2017 Minimum Air Temperature (°C)	2017 Maximum Air Temperature (°C)
1-Apr	14.2	7.3	21.5	15.8	6.7	25.8
2-Apr	12	5.2	21.9	14	7.7	21.4
3-Apr	11.7	8.1	18	14.3	9.1	22.3
4-Apr	16.5	8.3	25.4	15.1	9.1	22.8
5-Apr	22.9	14.2	30	16.6	9	25.2
6-Apr	21.9	13.1	33.2	15.5	9.8	21.2
7-Apr	15.5	11.4	21.7	14.9	13.1	16.1
8-Apr	14.7	13.4	16.7	13.1	9.3	16.5
9-Apr	15.9	12.7	20.9	13	8.8	20.9
10-Apr	15.2	12.5	20	15.7	10.2	23.2
11-Apr	14.9	12	20.5	14.7	9.2	22.4
12-Apr	13.8	9.6	19.4	15	10.1	20.5
13-Apr	14.7	9.2	21.1	13.7	8.4	17.5

Table 7. *Cont.*

Date	2016 Average Air Temperature (°C)	2016 Minimum Air Temperature (°C)	2016 Maximum Air Temperature (°C)	2017 Daily Average Air Temperature (°C)	2017 Minimum Air Temperature (°C)	2017 Maximum Air Temperature (°C)
14-Apr	13.1	6.6	19.1	13.7	8.7	21.7
15-Apr	14.9	10	20.1	16.6	10.3	25.5
16-Apr	17.3	9.2	26.1	15.2	8.6	22.5
17-Apr	18.6	7.7	30	15.3	12.6	18.7
18-Apr	17.9	7.4	29.6	14.6	12.2	17.7
19-Apr	17.6	8.7	28.5	16	10.8	21.4
20-Apr	15.5	5.9	25.6	17.1	10.4	23.9
21-Apr	14.9	7.2	22.6	18.6	12.7	27.5
22-Apr	14.3	7.7	21.2	17.1	11	26
23-Apr	14.5	7.8	20.2	15.8	11.1	22.3
24-Apr	14.3	6.6	20.2	15.1	11	20.4
25-Apr	10.5	3.4	16.8	14.8	11.5	19
26-Apr	11.8	5.1	19.9	16.2	13	23.7
27-Apr	12.7	6	19.5	16.2	11.1	22.5
28-Apr	11.5	5.2	17.7	16.4	11.7	23.9
29-Apr	12.1	8.3	18.5	18.3	10.6	26.2
30-Apr	14.1	7.3	21.3	18.1	9.5	27.7
1-May	15.3	7.8	23.1	23.6	15	31.8
2-May	14.5	11.5	20.7	23.9	16.8	31.7
3-May	14.3	10	23.1	21.5	14.6	27.1
4-May	14	11.4	19.4	16.8	12.3	26.7
5-May	15.5	12.9	19.8	14.5	10.5	23.1
6-May	14.7	9.9	20.3	11.4	8.4	15.1
7-May	14.8	9.7	20.5	11.3	9.2	15.4
8-May	14.3	7.6	20.9	13	7.8	19.1
9-May	15	12.3	20.2	14	9.4	19.8
10-May	14.6	11.9	20	14.8	13.2	17.6
11-May	15	12.3	20.2	15.7	11.4	21.1
12-May	14.6	11.7	20.3	14.6	10.4	20.8
13-May	14.1	11.6	18.8	13.4	9.1	18.8
14-May	15.3	8.5	21.4	12.4	7.4	17.2
15-May	15	10.5	20.1	12	7.9	16.6
16-May	14.2	9	20	13.1	9.6	17.9
17-May	15.7	10.7	22.4	13.1	9.5	18.2

Table 7. *Cont.*

Date	2016 Average Air Temperature (°C)	2016 Minimum Air Temperature (°C)	2016 Maximum Air Temperature (°C)	2017 Daily Average Air Temperature (°C)	2017 Minimum Air Temperature (°C)	2017 Maximum Air Temperature (°C)
18-May	16.3	9.9	25.3	16.9	9.3	27.1
19-May	13.5	11.5	19.7	21.5	11.7	29.2
20-May	13.3	9.2	16.9	19.6	10.9	29.4
21-May	12.9	6.2	18.1	17.9	10.5	27.9
22-May	13.3	5.9	19.4	16.2	10.3	24.6
23-May	14	6.9	19.8	16.3	11.7	24.2
24-May	14.6	11.9	19.1	15.3	11.6	23.6
25-May	14.7	11.6	18.5	15.3	13.8	18.5
26-May	14	8.4	18.8	15.5	10.8	19.9
27-May	15.1	10.5	21.5	15.2	8.2	21.5
28-May	15.4	10.6	21.8	15.9	10.6	21.5
29-May	15.3	12.6	21	15.7	12.7	20.8
30-May	15.6	12	22.4	15.7	11.9	22.4
31-May	16	10.4	24.6	17.4	11.4	24.7
1-Jun	16	11.2	23.4	17.9	12.8	26.7
2-Jun	16.2	10	27.1	19	13.1	29.2
3-Jun	19.4	9.5	29	18.6	12.4	26.3
4-Jun	18.8	11.9	32.4	17.4	10.6	25.4
5-Jun	15.9	10.3	23.1	16.8	10.7	26
6-Jun	15.9	11.9	22.3	16.7	10.7	24.7
7-Jun	15.9	12.3	22.8	16.7	12.5	23.1
8-Jun	16	11.8	22.1	17.4	9.7	24.9
9-Jun	15.2	11.4	20.4	18.1	14.5	23.4
10-Jun	16.6	11.5	24.7	16.6	10.1	23.8
11-Jun	16.3	10.7	21.8	15.5	9.6	23.5
12-Jun	17.5	11.6	22.2	16	8.8	23.3
13-Jun	15.7	11.4	21.4	17.7	12	26.5
14-Jun	15	8.7	21.8	21.1	14.7	28.8
15-Jun	13.5	7.4	18.7	23.5	14.6	30.3
16-Jun	16	8.4	23.4	24.7	18.8	31.9
17-Jun	16.8	9.9	24.8	26.7	18.3	32.3
18-Jun	18.5	12.4	27.1	23.3	17.9	29.5
19-Jun	22.3	12.6	31.9	21.8	13.5	32.5

Table 7. *Cont.*

Date	2016 Average Air Temperature (°C)	2016 Minimum Air Temperature (°C)	2016 Maximum Air Temperature (°C)	2017 Daily Average Air Temperature (°C)	2017 Minimum Air Temperature (°C)	2017 Maximum Air Temperature (°C)
20-Jun	23.6	14.5	33.6	18.6	13	29
21-Jun	24.3	14.3	32.9	17.9	12.7	26.1
22-Jun	17.5	10.3	27.2	19.2	11.8	29.1
23-Jun	21.6	9.7	34.3	18.9	15	26.1
24-Jun	19.4	11.6	25.9	18.3	14.1	24.7
25-Jun	21.7	12	34.3	18.3	13.2	26.4
26-Jun	23.6	13.9	34.4	19.5	10.7	30.4
27-Jun	26.7	17	35.3	18.4	13.4	28.3
28-Jun	22.4	12.5	30.7	17.8	13.1	25.6
29-Jun	16.8	11.4	24	17.4	12.7	26.7
30-Jun	14.7	10.2	21.9	17.3	12.4	27.4
1-Jul	16.5	11.6	23.1	16.8	12.3	23.5
2-Jul	17.3	12.6	23.4	18.2	13.3	25
3-Jul	16.3	11.6	23	19	12.9	26.4
4-Jul	15.7	10.2	21.7	18.6	10.8	28.2
5-Jul	16	13.1	22.7	19.5	13.9	31.5
6-Jul	15.6	12.2	22.1	22	15.3	30.3
7-Jul	15.7	12.3	21.8	26.1	16.5	35.9
8-Jul	16.1	11.7	24.1		17.6	38
9-Jul	18.2	11.9	28.7		12.6	28.7
10-Jul	17.2	11.2	26	20	14.5	30.8
11-Jul	16.3	10.9	23.1	21.9	14.7	29.8
12-Jul	16.9	11.3	26.9	21.6	12.7	27.8
13-Jul	16.2	9.9	24.4	19.3	13.6	28.4
14-Jul	17.1	11.1	26.4	18.6	13.5	27.2
15-Jul	16.1	10.6	25.4	18.3	13	31
16-Jul	16	12.3	22.5	20.5	16.7	33.7
17-Jul	16	11.3	23.2	23.9	14.6	29.8
18-Jul	16.7	10.5	24.2	21.8	14.1	28.6
19-Jul	15.9	9.2	24.8	20.5	11.9	27
20-Jul	18.1	10	27.6	17.9	10.8	27
21-Jul	17.6	10.9	28.4	18.9	11.8	28.1
22-Jul	22.1	13.5	31.9	18.5	11.9	27.7
23-Jul	22.8	13.9	33.2	19.8	13.3	30.5

Table 7. Cont.

Date	2016 Average Air Temperature (°C)	2016 Minimum Air Temperature (°C)	2016 Maximum Air Temperature (°C)	2017 Daily Average Air Temperature (°C)	2017 Minimum Air Temperature (°C)	2017 Maximum Air Temperature (°C)
24-Jul	19.3	11.5	28.8	18.7	14.2	25.6
25-Jul	16.9	8.7	26.6	18.9	15.1	24.8
26-Jul	19.5	9.5	29.7	18.9	13.8	26.2
27-Jul	19.9	13	28.2	19	14.5	27.5
28-Jul	17.8	12	26.4	18.7	13.5	25.8
29-Jul	17.6	12.2	26.8	18.1	14.1	26.4
30-Jul	18.7	11.9	27.7	17.8	14.2	25.3
31-Jul	18.6	13.2	26.9	18	13.5	26.4
1-Aug	18.4	13.8	26.9	20.1	14.2	28.8
2-Aug	17.8	14.1	24.9	22.6	17.8	29.6
3-Aug	17.4	13.6	24.8	23.8	19	31.1
4-Aug	16.4	13.6	22.8	22.9	16.8	31.7
5-Aug	15.7	13.4	20.7	19.9	15.9	29.2
6-Aug	16	13	22.2	19.4	15.6	26.2
7-Aug	16.3	11.6	22.7	19	15.1	26
8-Aug	15.5	10.4	24.2	18.7	14.4	26.6
9-Aug	15.4	10.6	21.5	19.1	15.4	26.4
10-Aug	16.9	13.7	23.2	19.1	15	28
11-Aug	16.9	12.5	23.8	19.2	13.2	28.9
12-Aug	16.2	11.6	23.7	19.1	13.4	33.4
13-Aug	17.5	12.5	25.9	18.6	13.6	29.1
14-Aug	17.6	12.4	27.5	18.1	13.9	29.2
15-Aug	16	11.9	23.6	17.9	13.7	22.2
16-Aug	15.9	11.5	24.2	18.2	13.1	24.7
17-Aug	16.8	11.6	26	19.3	14.2	27.2
18-Aug	16.2	11.6	24.4	19	13.9	25.6
19-Aug	16.3	12.5	23.6	19.4	15.6	27.1
20-Aug	16	12.4	22.6	19.3	16.8	24
21-Aug	16.2	12.9	22.6	20.2	16	26.2
22-Aug	16.7	13.3	23	19.7	13.7	27.3
23-Aug	16.7	13.7	22.7	19	16.5	24.3
24-Aug	15.9	13.2	21.7	19.3	15.7	29.1
25-Aug	16	13	22.6	20.2	13.9	29
26-Aug	16.5	13.5	22.3	20.5	14.8	29.6

Table 7. *Cont.*

Date	2016 Average Air Temperature (°C)	2016 Minimum Air Temperature (°C)	2016 Maximum Air Temperature (°C)	2017 Daily Average Air Temperature (°C)	2017 Minimum Air Temperature (°C)	2017 Maximum Air Temperature (°C)
27-Aug	17.2	13	21.7	20.8	14.7	30
28-Aug	17.1	13	22.8	19.9	13.7	30
29-Aug	18.1	12.4	28.3	20.1	15.5	28.1
30-Aug	20.5	13.5	32.7	21.1	14.5	30.5
31-Aug	17.2	11.6	24.7	22.6	16.2	31
1-Sep	15	10.9	21.1	28.7	18.5	39.1
2-Sep	16.1	10.2	23.7			
3-Sep	15.4	12.1	21.1	28.6	22.2	38.3
4-Sep	15.7	10.4	20.3	22.1	20.1	26
5-Sep	14.7	9.3	20.8	22.3	16.3	28.9
6-Sep	17.7	8.8	28.4	19.3	15.2	26.5
7-Sep	18.4	15.8	23.9	19.9	16.1	25.2
8-Sep	17.7	12.8	24.3	18.7	16.8	24.7
9-Sep	16	10.4	23	19.2	15.8	26.8
10-Sep	16.3	13.3	22.8	21.4	15.3	30.2
11-Sep	15.7	13.1	22	24.6	18.7	31.4
12-Sep	15.6	13.9	20.8	20.2	18	25.6
13-Sep	15.6	9.1	21.4	19.2	14.7	24.8
14-Sep	15	7.3	23.1	18.2	13.7	22.1
15-Sep	18.6	11	28.5	17	11.5	24.6
16-Sep	16	11.6	24.3	16.2	10.2	23.2
17-Sep	18.4	11.6	31.4	17.3	12	23.7
18-Sep	22.8	11.8	38.9	17.8	11.6	27.3
19-Sep	21.4	13.1	33.3	19.1	14.9	25.1
20-Sep	21.4	13.6	29.3	19.4	15.7	27.2
21-Sep	15.4	9.6	21.8	16	10	21.2
22-Sep	15.7	9.2	20.8	15.7	10	22.1
23-Sep	18.8	13.4	26.1	18.2	11.8	24.7
24-Sep	24.4	18.1	33.3	20.6	13.1	28.9
25-Sep	26.5	16.5	36.9	21.3	13	30.8
26-Sep	25.2	14.7	38.4	22.1	12.9	32.3
27-Sep	25.7	13.9	39.4	21	11.6	33.4
28-Sep	19	10.8	30.3	21.9	13.4	33.9

Table 7. *Cont.*

Date	2016 Average Air Temperature (°C)	2016 Minimum Air Temperature (°C)	2016 Maximum Air Temperature (°C)	2017 Daily Average Air Temperature (°C)	2017 Minimum Air Temperature (°C)	2017 Maximum Air Temperature (°C)
29-Sep	16.9	9	29.2	18.3	12.4	28.8
30-Sep	14.4	8.9	22.8	16.2	11.9	23.5
1-Oct	17.4	10.5	27.4	20.9	14.4	29.5
2-Oct	16.5	11.2	21.2	19.3	10.2	28.5
3-Oct	15.7	9.7	22.4	13.7	7.5	20
4-Oct	15.7	8.2	24.7	18.9	8.6	28.3
5-Oct	18	13.4	24.9	23.2	15.9	30.3
6-Oct	20.6	15	28.6	24	14.1	34.1
7-Oct	22.8	14.1	33.4	23.8	12.4	35.1
8-Oct	23.5	12.3	33	16.2	9.1	24.6
9-Oct	17.7	9.5	29.2	18.5	11.2	27.1
10-Oct	14.4	7	21.6	15.9	8.4	25.8
11-Oct	15	10	20.3	16.3	9	22.6
12-Oct	14.5	8.5	21.2	20.5	16.1	26.4
13-Oct	15.2	10.9	22.6	21.8	15.7	28.3
14-Oct	17.4	13.9	23.4	24.2	20.3	30.1
15-Oct	17.8	16.8	19.6	21.2	11.5	35.4
16-Oct	17.6	16.1	20.1	20.9	10.4	35
17-Oct	17.3	13.2	22	20.5	12.7	31.1
18-Oct	18.2	13.3	24.7	18.3	12.2	26.8
19-Oct	21.4	15.6	27.5	15.5	9.8	24.3
20-Oct	24.8	19.5	31.3	15.9	11.2	21.4
21-Oct	22.5	12.7	33	17.9	11.3	24.2
22-Oct	16.6	9.3	29.1	22.2	17.8	28.4
23-Oct	14.9	8	21.7	27.2	20.3	35.8
24-Oct	16.7	10.8	21.4		22.6	38.6
25-Oct	15.4	9.6	24.5	26.1	15.3	38
26-Oct	15.2	7.6	28	20.8	11.9	34.6
27-Oct	16.4	9.2	27.9	18.2	11	31.7
28-Oct	17.2	14.5	21.5	15.1	10.7	27.2
29-Oct	17.4	14	23.4	13.5	11.5	18
30-Oct	15.4	10.6	19.1	13.5	12.9	14.6
31-Oct	13.2	8	20.8	15.2	13	19

References

1. Harbertson, J.F.; Hodgins, R.E.; Thurston, L.N.; Schaffer, L.J.; Reid, M.S.; Landon, J.L.; Ross, C.F.; Adams, D.O. Variability of tannin concentration in red wines. *Am. J. Enol. Vitic.* **2008**, *59*, 210–214.
2. Dimitrovska, M.; Bocevska, M.; Dimitrovski, D.; Murkovic, M. Anthocyanin composition of Vranec, Cabernet Sauvignon, Merlot and Pinot Noir grapes as indicator of their varietal differentiation. *Eur. Food Res. Technol.* **2011**, *232*, 591–600. [CrossRef]
3. Waterhouse, A.L. Wine phenolics. *Ann. N. Y. Acad. Sci.* **2002**, *957*, 21–36. [CrossRef] [PubMed]
4. Boulton, R. The Copigmentation of Anthocyanins and Its Role in the Color of Red Wine: A Critical Review. *Am. J. Enol. Vitic.* **2001**, *52*, 67–87.
5. Jordheim, M.; Fossen, T.; Andersen, Ø.M. Characterization of Hemiacetal Forms of Anthocyanidin 3-*O* -β-Glycopyranosides. *J. Agric. Food Chem.* **2006**, *54*, 9340–9346. [CrossRef] [PubMed]
6. Bimpilas, A.; Panagopoulou, M.; Tsimogiannis, D.; Oreopoulou, V. Anthocyanin copigmentation and color of wine: The effect of naturally obtained hydroxycinnamic acids as cofactors. *Food Chem.* **2016**, *197*, 39–46. [CrossRef] [PubMed]
7. Poncet-Legrand, C.; Gautier, C.; Cheynier, V.; Imberty, A. Interactions between flavan-3-ols and poly(L-proline) studied by isothermal titration calorimetry: Effect of the tannin structure. *J. Agric. Food Chem.* **2007**, *55*, 9235–9240. [CrossRef] [PubMed]
8. Casassa, L.F.; Harbertson, J.F. Extraction, Evolution, and Sensory Impact of Phenolic Compounds during Red Wine Maceration. *Annu. Rev. Food Sci. Technol.* **2014**, *5*, 83–109. [CrossRef] [PubMed]
9. Lesschaeve, I.; Noble, A.C. Polyphenols: Factors influencing their sensory properties and their effects on food and beverage preferences. *Am. J. Clin. Nutr.* **2005**, *81*, 330S–335S. [CrossRef] [PubMed]
10. Harbertson, J.F.; Parpinello, G.P.; Heymann, H.; Downey, M.O. Impact of exogenous tannin additions on wine chemistry and wine sensory character. *Food Chem.* **2012**, *131*, 999–1008. [CrossRef]
11. Ferrer-Gallego, R.; Hernández-Hierro, J.M.; Rivas-Gonzalo, J.C.; Escribano-Bailón, M.T. Sensory evaluation of bitterness and astringency sub-qualities of wine phenolic compounds: Synergistic effect and modulation by aromas. *Food Res. Int.* **2014**, *62*, 1100–1107. [CrossRef]
12. Villamor, R.R.; Evans, M.A.; Mattinson, D.S.; Ross, C.F. Effects of ethanol, tannin and fructose on the headspace concentration and potential sensory significance of odorants in a model wine. *Food Res. Int.* **2013**, *50*, 38–45. [CrossRef]
13. Danilewicz, J. Review of Reaction Mechanisms of Oxygen and Proposed Intermediate Reduction Products in Wine: Central Role of Iron and Copper. *Am. J. Enol. Vitic.* **2003**, *54*, 73–85.
14. Fulcrand, H.; Dueñas, M.; Salas, E.; Cheynier, V. Phenolic reactions during winemaking and aging. *Am. J. Enol. Vitic.* **2006**, *57*, 289–297.
15. Cliff, M.A.; King, M.C.; Schlosser, J. Anthocyanin, phenolic composition, colour measurement and sensory analysis of BC commercial red wines. *Food Res. Int.* **2007**, *40*, 92–100. [CrossRef]
16. Van Buren, J.; Bertino, J. The stability of wine anthocyanins on exposure to heat and light. *Am. J. Enol. Vitic.* **1968**, *19*, 147–154.
17. Brouillard, R.; Chassaing, S.; Fougerousse, A. Why are grape/fresh wine anthocyanins so simple and why is it that red wine color lasts so long? *Phytochemistry* **2003**, *64*, 1179–1186. [CrossRef]
18. Somers, T.C.; Evans, M.E. Wine quality: Correlations with colour density and anthocyanin equilibria in a group of young red wines. *J. Sci. Food Agric.* **1974**, *25*, 1369–1379. [CrossRef]
19. Reynolds, A.G.; Price, S.F.; Wardle, D.A.; Watson, B.T. Fruit Environment and Crop Level Effects on Pinot-Noir. 1. Vine Performance and Fruit Composition in British-Columbia. *Am. J. Enol. Vitic.* **1994**, *45*, 452–459.
20. Guidoni, S.; Allara, P.; Schubert, A. Effect of cluster thinning on berry skin anthocyanin composition of *Vitis vinifera* cv. Nebbiolo. *Am. J. Enol. Vitic.* **2002**, *53*, 224–226.
21. Keller, M.; Mills, L.J.; Wample, R.L.; Spayd, S.E. Cluster Thinning Effects on Three Deficit-Irrigated *Vitis vinifera* Cultivars. *Am. J. Enol. Vitic.* **2005**, *56*, 91–103. [CrossRef]
22. Valdés, M.E.; Moreno, D.; Gamero, E.; Uriarte, D.; Prieto, M.D.H.; Manzano, R.; Picon, J.; Intrigliolo, D.S. Effects of cluster thinning and irrigation amount on water relations, growth, yield and fruit and wine composition of tempranillo grapes in extremadura (Spain). *J. Int. Sci. Vigne Vin* **2009**, *43*, 67–76.
23. Santesteban, L.G.; Miranda, C.; Royo, J.B. Thinning intensity and water regime affect the impact cluster thinning has on grape quality. *Vitis J. Grapevine Res.* **2011**, *50*, 159–165.

24. Gamero, E.; Moreno, D.; Talaverano, I.; Prieto, M.H.; Guerra, M.T.; Valdés, M.E. Effects of irrigation and cluster thinning on tempranillo grape and wine composition. *S. Afr. J. Enol. Vitic.* **2014**, *35*, 196–204. [CrossRef]

25. Uzes, D.; Skinkis, P.A. Factors Influencing Yield Management of Pinot Noir Vineyards in Oregon. *J. Ext.* **2016**, *54*, 2012–2014.

26. Palliotti, A.; Cartechini, A. Cluster thinning effects on yield and grape composition in different grapevine cultivars. *Acta Hortic.* **2000**, *512*, 111–119. [CrossRef]

27. Reynolds, A.G.; Schlosser, J.; Sorokowsky, D.; Roberts, R.; Willwerth, J.; De Savigny, C. Magnitude of viticultural and enological effects. II. Relative impacts of cluster thinning and yeast strain on composition and sensory attributes of Chardonnay Musqué. *Am. J. Enol. Vitic.* **2007**, *58*, 25–41. [CrossRef]

28. Bravdo, B.; Hepner, Y.; Loinger, C.; Cohen, S.; Tabacman, H. Effect of Crop Level and Crop Load on Growth, Yield, Must and Wine Composition, and Quality of Cabernet Sauvignon. *Am. J. Enol. Vitic.* **1985**, *36*, 125–131.

29. Jackson, D.I.; Lombard, P.B.; Kabinett, L.Q. Environmental and management practices affecting grape composition and wine quality—A review. *Am. J. Enol. Vitic.* **1993**, *44*, 409–430.

30. Kliewer, W.M.; Dokoozlian, N.K. Leaf area/crop weight ratios of grapevines: Influence on fruit composition and wine quality. *Am. J. Enol. Vitic.* **2005**, *56*, 170–181.

31. Ravaz, L. Sur la brunissure de la vigne. *C. R. Acad. Sci.* **1903**, *136*, 1276–1278.

32. Smart, R.E.; Dick, J.K.; Gravett, I.M.; Fisher, B.M. Canopy management to improve grape yield and wine quality principles and practices. *S. Afr. J. Enol. Vitic.* **1990**, *11*, 3–17. [CrossRef]

33. Frioni, T.; Zhuang, S.; Palliotti, A.; Sivilotti, P.; Falchi, R.; Sabbatini, P. Leaf removal and cluster thinning efficiencies are highly modulated by environmental conditions in cool climate viticulture. *Am. J. Enol. Vitic.* **2017**, *68*, 325–335. [CrossRef]

34. Reeve, A.L.; Skinkis, P.A.; Vance, A.J.; McLaughlin, K.R.; Tomasino, E.; Lee, J.; Tarara, J.M. Vineyard Floor Management and Cluster Thinning Inconsistently Affect 'Pinot noir' Crop Load, Berry Composition, and Wine Quality. *HortScience* **2018**, *53*, 318–328. [CrossRef]

35. Moreno Luna, L.H.; Reynolds, A.G.; Di Profio, F. Crop Level and Harvest Date Impact Composition of Four Ontario Winegrape Cultivars. I. Yield, Fruit, and Wine Composition. *Am. J. Enol. Vitic.* **2017**, *68*, 431–446. [CrossRef]

36. Howell, G.S. Sustainable grape productivity and the growth-yield relationship: A review. *Am. J. Enol. Vitic.* **2001**, *52*, 165–174.

37. Bergqvist, J.; Dokoozlian, N.; Ebisuda, N. Sunlight exposure and temperature effects on berry growth and composition of Cabernet Sauvignon and Grenache in the Central San Joaquin Valley of California. *Am. J. Enol. Vitic.* **2001**, *52*, 1–7.

38. Cañón, P.M.; González, Á.S.; Alcalde, J.A.; Bordeu, E. Composición fenólica del vino tinto: Efecto de chapoda de brotes y raleo de racimos. *Cienc. Investig. Agrar.* **2014**, *41*, 235–248. [CrossRef]

39. Gil-Muñoz, R.; Moreno-Pérez, A.; Vila-López, R.; Fernández-Fernández, J.I.; Martínez-Cutillas, A.; Gómez-Plaza, E. Influence of low temperature prefermentative techniques on chromatic and phenolic characteristics of Syrah and Cabernet Sauvignon wines. *Eur. Food Res. Technol.* **2009**, *228*, 777–788. [CrossRef]

40. Ough, C.S.; Nagaoka, R. Effect of Cluster Thinning and Vineyard Yields on Grape and Wine Composition and Wine Quality of Cabernet Sauvignon. *Am. J. Enol. Vitic.* **1984**, *35*, 30–34.

41. Freeman, B.M.; Kliewer, A.W.M. Effect of Irrigation, Crop Level and Potassium Fertilization on Carignane Vines. II. Grape and Wine Quality. *Am. J. Enol. Vitic.* **1983**, *34*, 197–207.

42. Dayer, S.; Prieto, J.A.; Galat, E.; Perez Peña, J. Carbohydrate reserve status of Malbec grapevines after several years of regulated deficit irrigation and crop load regulation. *Aust. J. Grape Wine Res.* **2013**, *19*, 422–430. [CrossRef]

43. Naor, A.; Gal, Y.; Bravdo, B. Crop load affects assimilation rate, stomatal conductance, stem water potential and water relations of field-grown Sauvignon blanc grapevines. *J. Exp. Bot.* **1997**, *48*, 1675–1680. [CrossRef]

44. Matthews, M.A.; Nuzzo, V. Berry size and yield paradigms on grapes and wines quality. *Acta Hortic.* **2007**, *754*, 423–436. [CrossRef]

45. *California Grape Acreage Report 2012*; California Department of Food and Agriculture: Sacramento, CA, USA, 2013. Available online: https://www.nass.usda.gov/Statistics_by_State/California/Publications/Specialty_and_Other_Releases/Grapes/Acreage/2013/201303grpac.pdf (accessed on 16 June 2018).

46. Settevendemie, M. *San Luis Obispo County Department of Agriculture/Weights and Measures Annual Report*; County Government Center: San Luis Obispo, CA, USA, 2016.

47. Coombe, B. Growth Stages of the Grapevine: Adoption of a system for identifying grapevine growth stages. *Aust. J. Grape Wine Res.* **1995**, *1*, 104–110. [CrossRef]

48. Winkler, A.J.; Cook, J.A.; Kliewer, W.M.; Lider, L.A. General Viticulture. *Soil Sci.* **1975**, *120*, 462. [CrossRef]

49. Harbertson, J.F.; Picciotto, E.A.; Adams, D.O. Measurement of Polymeric Pigments in Grape Berry Extracts and Wines Using a Protein Precipitation Assay Combined with Bisulfite Bleaching. *Am. J. Enol. Vitic.* **2003**, *54*, 301–306.

50. Harbertson, J.F.; Kennedy, J.A.; Adams, D.O. Tannin in Skins and Seeds of Cabernet Sauvignon, Syrah, and Pinot noir Berries during Ripening. *Am. J. Enol. Vitic.* **2002**, *53*, 54–59.

51. Meilgaard, M.; Carr, B.; Civille, G. *Sensory Evaluation Techniques*; CRC Press: Boca Raton, FL, USA, 2006; p. 388.

52. Kliewer, W.M. Berry composition of *Vitis vinifera* cultivars as influenced by photo- and nycto-temperatures during maturation. *J. Am. Soc. Hortic. Sci.* **1973**, *98*, 153–159.

53. Reeve, A.L.; Skinkis, P.A.; Vance, A.J.; Lee, J.; Tarara, J.M. Vineyard floor management influences "Pinot noir" vine growth and productivity more than cluster thinning. *HortScience* **2016**, *51*, 1233–1244. [CrossRef]

54. McIntyre, G.N.; Kliewer, W.M.; Lider, L.A. Some Limitations of the Degree Day System as Used in Viticulture in California. *Am. J. Enol. Vitic.* **1987**, *38*, 128–132.

55. Dokoozlian, N.K.; Kliewer, W.M. Influence of Light on Grape Berry Growth and Composition Varies during Fruit Development. *J. Am. Soc. Hortic. Sci.* **1996**, *121*, 869–874.

56. Price, S.F.; Breen, P.J.; Valladao, M.; Watson, B.T. Cluster Sun Exposure and Quercetin in Pinot noir Grapes and Wine. *Am. J. Enol. Vitic.* **1995**, *46*, 187–194.

57. Rienth, M.; Torregrosa, L.; Sarah, G.; Ardisson, M.; Brillouet, J.; Romieu, C. Temperature desynchronizes sugar and organic acid metabolism in ripening grapevine fruits and remodels their transcriptome. *BMC Plant Biol.* **2016**, *16*, 164. [CrossRef] [PubMed]

58. Boban, N.; Tonkic, M.; Budimir, D.; Modun, D.; Sutlovic, D.; Punda-Polic, V.; Boban, M. Antimicrobial Effects of Wine: Separating the Role of Polyphenols, pH, Ethanol, and Other Wine Components. *J. Food Sci.* **2010**, *75*, M322–M326. [CrossRef] [PubMed]

59. Singleton, V.L. Oxygen with Phenols and Related Reactions in Musts, Wines, and Model Systems: Observations and Practical Implications. *Am. J. Enol. Vitic.* **1987**, *38*, 69–77.

60. Guerzoni, M.E.; Sinigaglia, M.; Gardini, F.; Ferruzzi, M.; Torriani, S. Effects of pH, Temperature, Ethanol, and Malate Concentration on *Lactobacillus plantarum* and *Leuconostoc oenos*: Modelling of the Malolactic Activity. *Am. J. Enol. Vitic.* **1995**, *46*, 368–374.

61. Ough, C.S.; Amerine, M.A. Studies with Controlled Fermentation X. Effect of Fermentation Temperature on Some Volatile Compounds in Wine. *Am. J. Enol. Vitic.* **1967**, *18*, 157–164.

62. Du, G.; Zhan, J.; Li, J.; You, Y.; Zhao, Y.; Huang, W. Effect of Fermentation Temperature and Culture Medium on Glycerol and Ethanol during Wine Fermentation. *Am. J. Enol. Vitic.* **2012**, *63*. [CrossRef]

63. Coombe, B.G. The Development of Fleshy Fruits. *Annu. Rev. Plant Physiol.* **1976**, *27*, 207–228. [CrossRef]

64. Kliewer, W.M. Effect of High Temperatures during the Bloom-Set Period on Fruit-Set, Ovule Fertility, and Berry Growth of Several Grape Cultivars. *Am. J. Enol. Vitic.* **1977**, *28*, 215–222.

65. Gillaspy, G. Fruits: A Developmental Perspective. *Plant Cell Online* **1993**, *5*, 1439–1451. [CrossRef] [PubMed]

66. Van Volkenburgh, E. Leaf expansion—An integrating plant behaviour. *Plant Cell Environ.* **1999**, *22*, 1463–1473. [CrossRef]

67. Mullins, M.G.; Bouquet, A.; Williams, L.E. *Biology of the Grapevine*; Cambridge University Press: Cambridge, UK, 1992; ISBN 0-521-30507-1.

68. Jurd, L. Review of Polyphenol Condensation Reactions and their Possible Occurrence in the Aging of Wines. *Am. J. Enol. Vitic.* **1969**, *20*, 191–195.

69. Chris Somers, T.; Evans, M.E. Spectral evaluation of young red wines: Anthocyanin equilibria, total phenolics, free and molecular SO2, "chemical age". *J. Sci. Food Agric.* **1977**, *28*, 279–287. [CrossRef]

70. Casassa, L.F. Flavonoid Phenolics in Red Winemaking. In *Phenolic Compunds—Natural Sources, Importance and Applications*; Soto-Hernandez, M., Ed.; 2017; pp. 153–196. Available online: https://www.intechopen.com/books/phenolic-compounds-natural-sources-importance-and-applications/flavonoid-phenolics-in-red-winemaking (accessed on 6 June 2018).

71. Heredia, F.J.; Francia-Aricha, E.M.; Rivas-Gonzalo, J.C.; Vicario, I.M.; Santos-Buelga, C. Chromatic characterization of anthocyanins from red grapes—I. pH effect. *Food Chem.* **1998**, *63*, 491–498. [CrossRef]
72. Weber, F.; Greve, K.; Durner, D.; Fischer, U.; Winterhalter, P. Sensory and Chemical Characterization of Phenolic Polymers from Red Wine Obtained by Gel Permeation Chromatography. *Am. J. Enol. Vitic.* **2013**, *64*, 15–25. [CrossRef]
73. Casassa, L.F.; Larsen, R.C.; Beaver, C.W.; Mireles, M.S.; Keller, M.; Riley, W.R.; Smithyman, R.; Harbertson, J.F. Sensory Impact of Extended Maceration and Regulated Deficit Irrigation on Washington State Cabernet Sauvignon Wines. *Am. J. Enol. Vitic.* **2013**, *64*, 505–514. [CrossRef]
74. Girard, B.; Kopp, T.G.; Reynolds, A.G.; Cliff, M. Influence of vinification treatments on aroma constituents and sensory descriptors of Pinot noir wines. *Am. J. Enol. Vitic.* **1997**, *48*, 198–206.
75. Bubola, M.; Sivilotti, P.; Poni, S. Early leaf removal has a larger effect than cluster thinning on grape phenolic composition in cv. teran grape phenolic composition. *Am. J. Enol. Vitic.* **2017**, *68*, 234–242. [CrossRef]

![fermentation logo] *fermentation*

MDPI

Article

End-User Software for Efficient Sensor Placement in Jacketed Wine Tanks

Dominik Schmidt [1,*]**, Maximilian Freund** [2] **and Kai Velten** [1]

[1] Department of Modeling and Systems Analysis, Hochschule Geisenheim University, Von-Lade-Straße 1, 65366 Geisenheim, Germany; kai.velten@hs-gm.de

[2] Department of Enology, Hochschule Geisenheim University, Von-Lade-Straße 1, 65366 Geisenheim, Germany; maximilian.freund@hs-gm.de

* Correspondence: dominik.schmidt@hs-gm.de; Tel.: +49-6722-502-79734

Received: 24 May 2018; Accepted: 6 June 2018; Published: 9 June 2018

Abstract: In food processing, temperature is a key parameter affecting product quality and energy consumption. The efficiency of temperature control depends on the data provided by sensors installed in the production device. In the wine industry, temperature sensor placement inside the tanks is usually predetermined by the tank manufacturers. Winemakers rely on these measurements and configure their temperature control accordingly, not knowing whether the monitored values really represent the wine's bulk temperature. To address this problem, we developed an end-user software which 1. allows winemakers or tank manufacturers to identify optimal sensor locations for customizable tank geometries and 2. allows for comparisons between actual and optimal sensor placements. The analysis is based on numerical simulations of a user-defined cooling scenario. Case studies involving two different tanks showed good agreement between experimental data and simulations. Implemented based on the scientific Linux operating system gmlinux, the application solely relies on open-source software that is available free of charge.

Keywords: wine; temperature control; sensor placement; CFD; end-user software

---

1. Introduction

Temperature is an important factor in various processing steps during wine making [1]. Prior to fermentation, for example, a speed-up of must clarification by particle sedimentation can be achieved by appropriate tank cooling. During fermentation, temperature control allows the winemaker to define the aromatic profile of the wine and to prevent stuck or sluggish fermentations [2–6]. Other process steps involving temperature control are cold stabilization, malolactic fermentation or storage (aging) [1,7–10]. During active fermentation, a homogeneous temperature in the tank is achieved by efficient bubble mixing, while in processes where mixing is driven entirely by natural convection, temperature gradients in the tank are more likely [11–16].

To monitor wine temperatures, tanks are usually equipped with a single temperature sensor which is beneficial not only in terms of costs, but also regarding issues of hygienic design and clean-in-place installations. Usually, tank manufacturers do not reveal any details of their strategies and considerations behind the placement of that single sensor.

According to one manufacturer, for tanks with diameters in the range of 0.82 m to 3.6 m, winemaker's may choose from a few different positions along the tank wall in combination with one or two different distances from the wall.

Cooling applications for wine tanks include various types of double jackets as well as plates or coils that can be immersed into the wine as required. Typically, modern tanks are equipped with a pillow plate double jacket and can be ordered fully insulated [1,17]. For these tank types, it is a common practice to keep some distance between the sensor and the tank wall where sensor values might be biased by the double-jacket cooling.

Effective temperature control is important for wine quality, e.g., by hindering evaporative losses or degradation of aromatic compounds triggered by inappropriate temperatures during storage or aging, but it is also important in terms of overall energy costs since ill-positioned sensors indicating too high bulk temperatures may obviously lead to unnecessarily high cooling loads. In fact, heating and cooling applications have been identified as the main contributors to the energy costs of industrial winerys [8,18–22].

Recent studies on sensors in winemaking focus on developing smart monitoring systems in the fields of wireless sensor networks and Internet of Things (IoT) not covering the topic where temperature probes should be located [23–28]. Hence, this study aims on addressing the question on efficient sensor placement in winemaking for the first time.

Considering the broad range of available tank geometries and possible sensor positions, any experiment-based optimal sensor placement is obviously inefficient and infeasible. Hence, a procedure based on numerical simulations and computational fluid dynamics (CFD) is suggested in this study. The approach was implemented into an end-user software which offers practitioners a tool for the analysis of their specific configurations. It is limited to scenarios where the tank is filled with a liquid phase only, as scenarios with a large amount of particular matter, e.g., red wine mash, would need a different modeling approach.

2. Materials and Methods

2.1. Mathematical Model

To identify suitable temperature sensor locations inside wine fermentation tanks, the software computes simulations of cooling scenarios. Before and after fermentation, the flow is assumed to be driven by natural convection only. Density variations caused by typical wine cooling conditions are assumed small, s.t. the Boussinesq approximation can be used [1,29]. Under these assumptions, numerical simulations can be performed using the buoyantBoussinesqPimpleFoam solver of the open-source C++-CFD-toolbox OpenFOAM® [30].The underlying model describes single phase flow of an incompressible fluid where density variations are only accounted for in the buoyancy term following the Boussinesq approach.

Balance Equations

Assuming laminar flow, Refs. [31,32] the following system of partial differential equations for mass (Equation (1)), momentum (Equation (2)) and energy is applicable (Equation (3)):

$$\nabla \cdot u = 0, \tag{1}$$

$$\frac{\partial u}{\partial t} + \nabla \cdot (uu) = -\nabla \left(\frac{p}{\rho} \right) + \nabla \cdot (\nu \nabla u) + g_b, \tag{2}$$

$$\frac{\partial T}{\partial t} + \nabla \cdot (Tu) = \nabla \cdot \left(\frac{\nu}{\mathrm{Pr}} \nabla T \right), \tag{3}$$

with u, p, ρ, g_b, ν, T and Pr representing velocity ($\mathrm{m\,s^{-1}}$), pressure (Pa), density ($\mathrm{kg\,m^{-3}}$), Boussinesq gravity ($\mathrm{m\,s^{-2}}$), kinematic viscosity ($\mathrm{m^2\,s^{-1}}$), temperature (K) and Prandtl number (-). The Prandtl number is calculated as $\mathrm{Pr} = \frac{c_p \rho \nu}{k}$ with the specific heat capacity c_p ($\mathrm{J\,kg^{-1}\,K^{-1}}$).

Applying Boussinesq's approximation, the gravity term g_b is expressed as

$$g_b = [1.0 - \beta(T - T_{\mathrm{Ref}})] \, g, \tag{4}$$

with the thermal expansion coefficient β ($\mathrm{K^{-1}}$) and gravity g ($\mathrm{m\,s^{-2}}$).

2.2. Software Implementation

The end-user software is based on open-source software packages, including R (v.3.4.0) [33], Shiny by RStudio (v.1.0.3) [34], Salome (v.7.8.0) [35], OpenFOAM® (v.4.1) [30] and ParaView (v.5.0.1) [36]. To facilitate its use, it was implemented based on the scientific Linux operating system gmlinux (v.17.01) [37,38], which is an extended version of Ubuntu 16.04 LTS that provides all necessary programs pre-installed and pre-configured.Additional R packages—ggplot2 (v.2.2.1.9000) [39], Cairo (v.1.5-9) [40] and future (v.1.5.0) [41]—are installed automatically when launching the application for the first time.Computer intensive calculations such as meshing and numerical simulation are automatically run in parallel on all but one of the available processor cores. This ensures quick results while maintaining the system's operability.

The graphical user interface is based on Shiny. It is divided into two main panels. On the left panel, data input is realized using sliders thematically distributed over four tabs, *Tank*, *Cooling*, *Wine* and *Simulation*. More details on the content of these tabs are given in Section 2.3. The right panel is used for the output of responsive graphics and texts such as tank geometry sketches or information on the expected cooling rate. A progress bar is shown when the analysis is ran in the background. In case of invalid input, data warning messages are displayed. A screenshot of the user interface is shown in Figure 1.

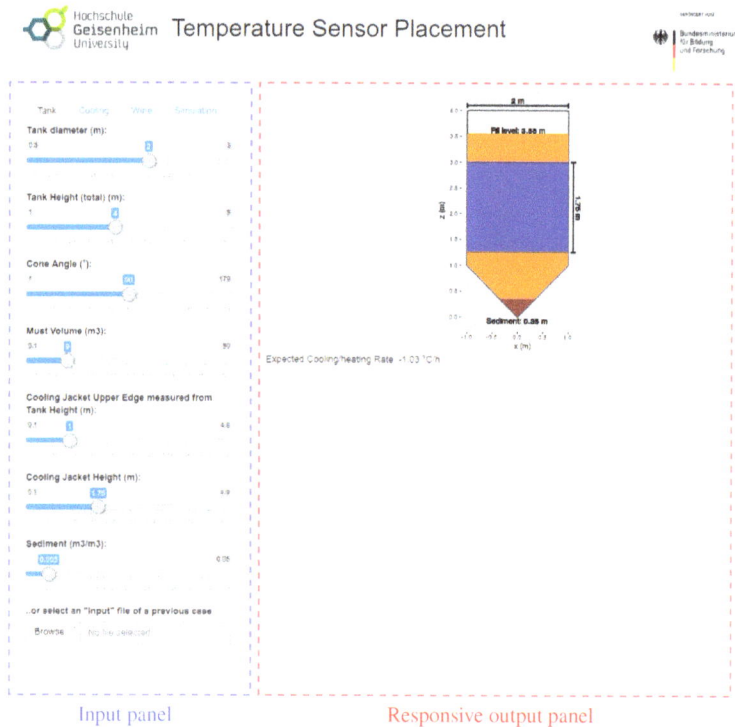

Figure 1. Screenshot of the graphical user interface. Dashed boxes indicate the input (**left**) and the responsive information panel (**right**).

2.3. Case Definition

The case setup in the software environment is organized in three steps requiring user input. In the first step, geometry data of the tank, dimensions of the cooling jacket and a fill level must be supplied. This includes tank diameter d_T (m), total tank height h_T (m), cone angle φ (°), liquid volume V (m^3), sediment fraction f_{Sed} (m^3 m^{-3}), cooling jacket distance from tank height $h_{0,J}$ (m) and jacket height h_J (m). In the second step, temperature conditions and cooling power of the scenario are defined. Here, the initial liquid temperature $T_{0,l}$ (°C) and the cooling power $\dot{Q}_c$ (W) are required. In the case of non-insulated tanks, the heat flow rate from ambient air to the liquid, $\dot{Q}_a$ (W), can also be set. To support user input, the expected cooling rate is calculated on the fly by Equation (5) and printed out in the right panel.

$$\dot{T}_c = \frac{(\dot{Q}_c + \dot{Q}_a) \cdot 3600\,\text{s}}{\rho V c_p} \quad (°\text{C h}^{-1}).$$

(5)

Thermo-physical properties of the liquid can be modified as required. Default values refer to a typical white wine after fermentation and are given in Table 1.

Table 1. Required thermo-physical properties of the liquid and their default values.

Variable	Value
c_p	$3960\,\text{J kg}^{-1}\,\text{K}^{-1}$
ρ	$970\,\text{kg m}^{-3}$
ν	$1.6 \times 10^{-6}\,\text{m}^2\,\text{s}^{-1}$
β	$207 \times 10^{-6}\,\text{K}^{-1}$
κ	$0.6\,\text{W m}^{-1}\,\text{K}^{-1}$

In the last step, model parameters relating to the optimal sensor placement procedure are defined. This includes simulation time t_{sim} (s), e.g., the length of a cooling cycle, the measurement interval of the temperature sensor Δt_{sens} (s), the accepted temperature tolerance ΔT_{sens}, e.g., the sensor's measurement tolerance, and the minimum percentage of valid measurements q (%). The simulation setup also allows for a choice of grid cell size Δx (mm) which can be used for a trade-off between level of uncertainty and computational costs. If required, input data from previous analyses can be read from a file. Further details on the use of the input parameters and on the simulation setup are provided in the following sections.

2.4. Geometry and Mesh Generation

The software visualizes the geometrical parameters provided by the user as a 2D-sketch of the tank including fill level and sediment volume. This is useful for a fine-tuning of geometrical parameters until they match real life situations such as truncated cone bottoms. Although the current implementation supports cylindro-conical tanks only, it can also be used to approximate dished bottoms as explained in Section 2.7. Figure 2 shows the default geometrical input data and the corresponding output sketch.

The software encompasses a fully automated mesh generation procedure (Figure 3) solely based on the geometrical input parameters and the grid cell size set by the user. Using a Python script, these data are used to generate a CAD-Model in Salome. This is realized using a rotational geometry approach referring to the outline of the fluid region of the tank as well as on information on cooling jacket position. The resulting STL-surfaces already represent the boundary patches used for the simulation setup as explained in Section 2.5. Finally, the *snappyHexMesh* utility of OpenFOAM® is used to built a hexahedral-dominant mesh, starting with an underlying *blockMesh* that uses the previously defined grid cell size Δx.

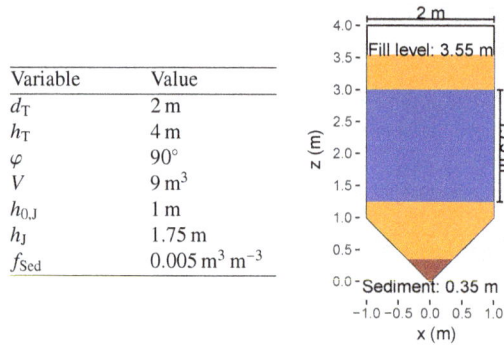

Figure 2. Exemplary geometrical input data with corresponding 2D tank sketch.

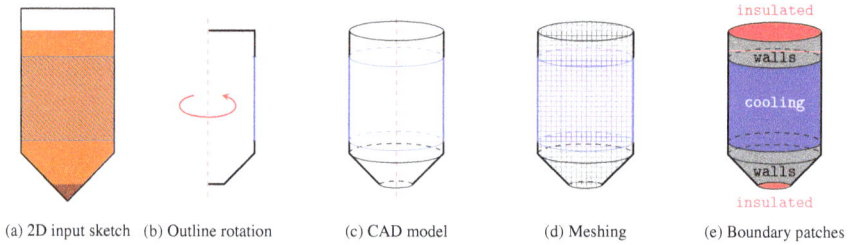

(a) 2D input sketch (b) Outline rotation (c) CAD model (d) Meshing (e) Boundary patches

Figure 3. Five steps (**a–e**) of the fully automated mesh generation workflow.

2.5. Boundary Conditions

The computational mesh is divided into three groups of boundary patches referred to as walls, cooling and insulated (Figure 3). The walls patches represent all tank surfaces in contact with liquid and ambiance where the specified heat flow is applied. This excludes the tank bottom where sediment is assumed to insulate the fluid from ambiance, and the liquid surface where insulation caused by the air in the tank's headspace is assumed. These regions are treated as a part of the insulated patch where a zero gradient Neumann-type boundary conditions is applied for temperature. The cooling patch corresponds to the jacketed tank surface. To express rates of heat flow ($\dot{Q}$) from the cooling jacket or from ambient air, Neumann-type boundary condition are applied. Boundary face values are evaluated using

$$T_f = T_x + \Delta_{fx} \nabla T_{lref} \, , \tag{6}$$

where T_x (K) is the current cell center temperature and Δ_{fx} (m) is the face-to-cell distance.

The local reference temperature gradient ∇T_{lref} is calculated from the rate of heat flow $\dot{Q}$ as follows:

$$\nabla T_{lref} = \frac{\dot{Q}}{A_P \cdot \rho c_P \alpha_{eff}} \, , \tag{7}$$

where A_P and α_{eff} describe the boundary patch area (m^2) and the effective thermal diffusivity (m^2 s^{-1}). Effective thermal diffusivity is calculated according to

$$\alpha_{eff} = \frac{\nu}{Pr} + \frac{\nu_t}{Pr_t} \, , \tag{8}$$

which simplifies to

$$\alpha_{eff} \equiv \alpha = \frac{\nu}{Pr} = \frac{k}{\rho c_P} \, , \tag{9}$$

since $v_t = 0$ is assumed.

A no-slip Dirichlet-type boundary conditions is applied for the velocity on all patches. Pressure is handled by OpenFOAM®'s *fixedFluxPressure* boundary condition, which acts similar to a zero gradient Neumann-type boundary condition including adjustments for body forces like gravity [42].

2.6. Identification of Sensor Locations

The sensor location algorithm identifies locations where a temperature sensor would most likely report the current liquid's bulk temperature $\bar{T}$ (K), ideally at all times. To guide the algorithm, the user is allowed to choose a threshold for acceptable temperature deviations, a measurement interval and a percentage of valid measurements; this can e.g., be used to better represent practical situations or the requirements discussed in Section 2.3. Based on the input data for these three constraints, most reliable temperature sensor locations are determined as follows. Each grid cell center is used as a potential sensor location. The user-defined measurement interval t_{sens} and overall simulation duration t_{sim} define the total number of sensor evaluations. To evaluate the sensor in a particular grid cell center, the deviation of the cell's temperature from the bulk mean temperature is determined and compared to the user-defined temperature threshold ΔT_{sens}. The overall percentage of valid measurements is calculated for every cell according to

$$q_x = \frac{1}{t_{sim}/\Delta t_{sens}} \sum_{1}^{t_{sim}/\Delta t_{sens}} \Theta\left(\Delta T_{sens} - |(\bar{T} - T_x)|\right), \tag{10}$$

with the Heaviside step function defined as:

$$\Theta: \mathbb{R} \to \{0,1\},$$
$$x \mapsto \begin{cases} 0: & x < 0 \\ 1: & x \geq 0. \end{cases} \tag{11}$$

All grid cells exceeding the user-defined threshold value q that defines the desired minimum percentage of valid measurements are classified as reliable sensor locations. An automated 3D ParaView visualization procedure is used to display the recommended sensor location cells along with a sketch of the tank geometry.

2.7. Case Studies

A proof of concept study was performed to evaluate the effectiveness and robustness of the sensor location algorithm. Two independent experiments were carried out to compare the software's simulation results with measured temperature profiles referring to wine tank cooling. Please note that the validity of the numerical solver *buoyantBoussinesqPimpleFoam* has already been proven elsewhere [43,44]. Insulated wine tanks equipped with a pillow plate cooling jacket were filled with an appropriate amount of tap water. After a 24 h resting period, initial bulk temperature was determined. Then, coolant was pumped through the cooling jacket for 1 h while monitoring the liquid's temperature at six different locations. At the end of this cooling cycle, the tank's content was homogenized using a mechanical mixer which resulted in a homogeneous, final bulk temperature. The difference between initial and final bulk temperatures was used to compute the cooling power required in the computer simulation of the experiment.

This experiment was carried out twice using two different-sized dished bottom wine tanks, both fully insulated and equipped with pillow plate cooling jackets. Both tanks were placed on load cells to measure the tanks' contents. The smaller tank had a diameter of $d_{T,S} = 0.6$ m and was filled with 300 kg of water, the larger tank ($d_{T,L} = 1.4$ m) was filled with 5420 kg. Temperature measurements were realized through thermowells on three different heights and two different depths using external probe temperature sensors (TSN-EXT44, AREXX Engineering, Zwolle, NL). With an accuracy of ±0.5 °C to

±1 °C on an operating temperature range from −30 °C to 80 °C, these sensors are similar to standard built-in wine tank sensors. The two sensors depths, measured as distances from the tank wall, for the small and the large tank were 5.5 cm/13 cm and 21 cm/34 cm, respectively. Details on the different measurement heights and tank dimensions are given in Figure 4. In the following, the six sensor locations are referred to as T-N, T-F, C-N, C-F, B-N and B-F, where T, C and B define the heights (top, center, bottom) and wall distance is expressed as N and F (near, far).

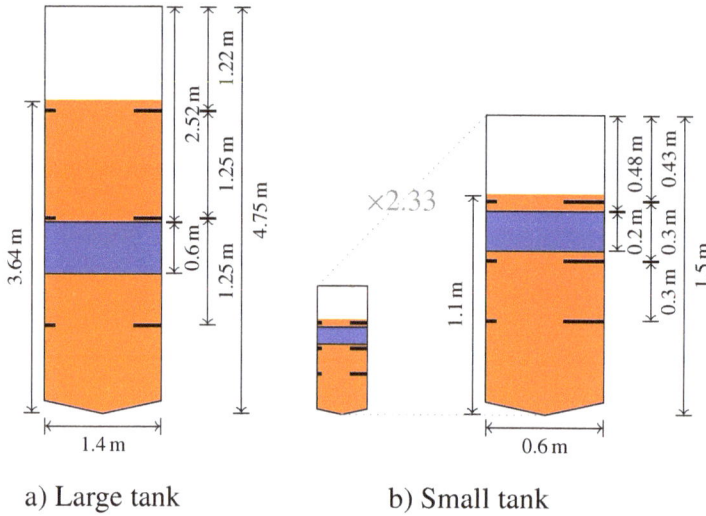

a) Large tank   b) Small tank

Figure 4. Dimensions, cooling jacket configuration and sensor locations of the (**a**) large and (**b**) small case study tank. To compare sizes the large tank and left version of the small tank are drawn on the same scale.

We measured an initial bulk temperature ($T_{0,1}$) of 13.4 °C for the large, 18.2 °C for the small tank. Final bulk temperatures were 12.92 °C in the large tank ($\Delta T = 0.48$ °C), and 16.88 °C in the small tank ($\Delta T = 1.32$ °C). This includes a correction for the rate of heat flow introduced by the mechanical mixer (TS-V17 movable mixer, 0.75 kW, 1400 rpm—Theo & Klaus Schneider GmbH & Co KG, Bretzenheim, Germany) of 405 W. Hence, for the large tank the applied cooling power ($\dot{Q}_c$) was calculated to be −3045 W from

$$\dot{Q}_c = mc_P \Delta T t_c^{-1}, \tag{12}$$

with $c_P = 4180 \, \text{J kg}^{-1} \text{K}^{-1}$ and $t_c = 3600 \, \text{s}$. In the experiment with the small tank, the cooling power was −460 W.

These experiments were then simulated in the software environment using the input data shown in Table 2. The values for kinematic viscosity, thermal expansion and thermal conductivity were taken from literature and set to $\nu = 1.5 \times 10^{-6} \, \text{m}^2 \text{s}^{-1}$, $\beta = 207 \times 10^{-6} \, \text{K}^{-1}$ and $\kappa = 0.56 \, \text{W m}^{-1} \text{K}^{-1}$ [45,46]. In addition to the standard software procedure, the simulation data was probed at the experimental sensor positions to gather temperature progressions curves similar to the experimental setup. To account for temperature variations in rotationally symmetric positions, radial mean temperatures were computed from the simulation results for each sensor position.

Table 2. Case studies: software input data for the small and large tank geometry.

Variable	Small Tank	Large Tank
d_T	0.6 m	1.4 m
h_T	1.5 m	4.75 m
φ	156°	156°
V_1	0.3 m^3	5.42 m^3
$h_{0,J}$	0.48 m	2.52 m
h_J	0.2 m	0.6 m
f_{Sed}	0.005 m^3 m^{-3}	0.005 m^3 m^{-3}

3. Results and Discussion

3.1. Case Studies

In the case studies, experimental temperature curves obtained at six different locations during a cooling cycle of 1 h were compared with simulation results. Initially, sensor data were calibrated to match initial bulk temperature. An exception was made for the B-N sensor in the large tank since its data showed inconsistencies during the first minutes. Its initial value was adjusted to represent a more physical temperature development.

In the large tank, only the lower region represented by sensors B-F and B-N was affected during 1 h of cooling (Figure 5), which can be explained by the fact that the cooling jacket was deeply immersed below the water surface (1.41 m). Minor differences in the temperature curves reported by B-F and B-B are within the range of sensor data fluctuations and thus insignificant. All remaining sensors including those located slightly above the cooling jacket (C-F, C-N) did not measure any cooling effect.

These results could be reproduced in simulations based on the end-user software and parameter settings described above. Two different grid resolutions, $\Delta x = 30$ mm and 20 mm, were used and gave similar results. A small cooling effect was detected in the simulations at sensor positions C-F and C-N, starting after approx. 30 min. Since this effect was smaller than the sensors tolerance limit, it was not observed in the experiments. We also noted that the temperature drop was more pronounced in the coarser grid case which indicates artificial numerical diffusion effects [29]. As Figure 5 shows, none of the large tank sensor locations under investigation was suitable to monitor the liquid's bulk temperature based on a ±0.1 °C tolerance level.

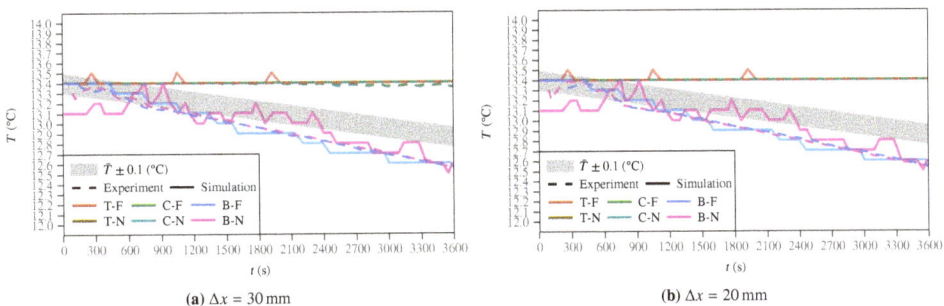

(a) $\Delta x = 30$ mm

(b) $\Delta x = 20$ mm

Figure 5. Comparison of experimental cooling data in the large tank with simulations, referring to six different sensor locations inside the tank and grid cell sizes of 30 mm (**a**) and 20 mm (**b**).

In the second case study with the smaller tank, the cooling jacket was located approximately 8 cm below the liquid surface. Similar to the large tank results, the sensors above the cooling jacket (T-F and T-N) did not detect a temperature drop during 1 h of cooling (Figure 6). All four sensors below the cooling jacket reported a steady temperature decrease while slightly lower temperatures were found closer to the tank bottom (sensors B-F and B-N). This corresponds to a typical natural convection flow pattern where cold fluid with higher density accumulates at the bottom. For the smaller tank, experimental data were compared with simulations on three different grid resolutions Δx: 30 mm, 20 mm and 10 mm. A good coincidence between simulations and experiments was found for all grid resolutions (Figure 6). Refining the grid resulted in a slightly larger temperature difference between the two lower sensor levels C-F/N and B-F/N without affecting the general cooling pattern.

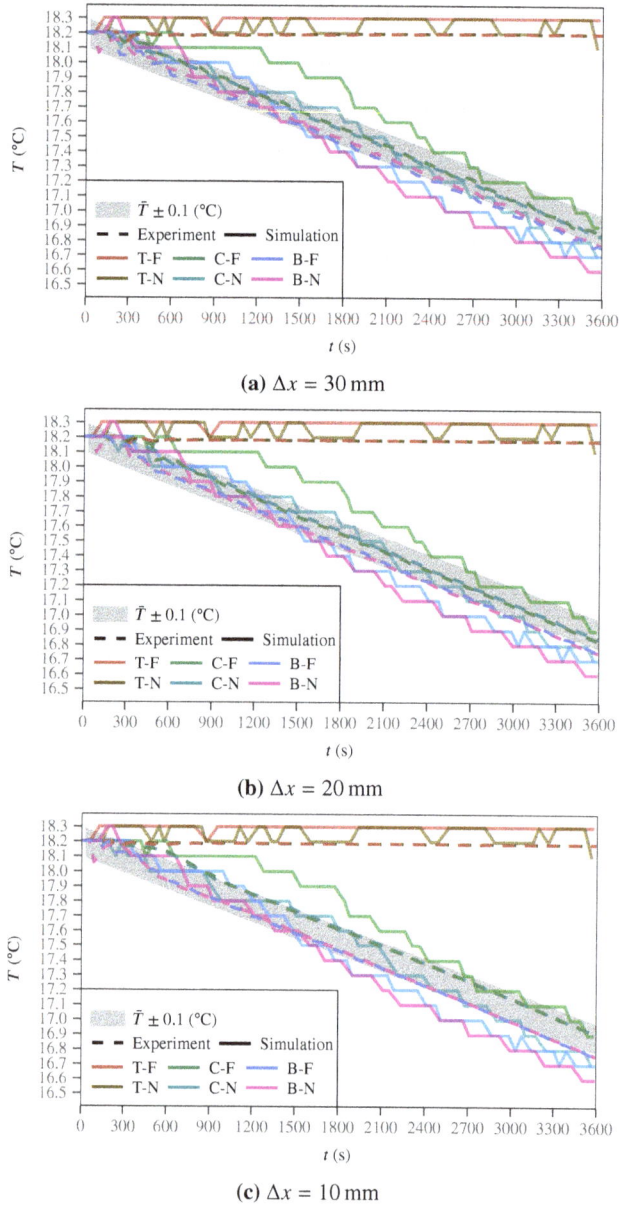

(a) $\Delta x = 30$ mm

(b) $\Delta x = 20$ mm

(c) $\Delta x = 10$ mm

Figure 6. Comparison of the experimental cooling process in the small tank with simulation data on three different grid resolutions (Δx), 30 mm (**a**); 20 mm (**b**) and 10 mm (**c**), referring to six different sensor locations inside the tank.

The simulation results indicated that sensor positions C-F and C-N seemed preferable to monitor average bulk temperature. This will be further investigated in the following section where the consequences of different settings of the accepted percentage of valid measurements (q) and the temperature tolerance (ΔT_{sens}) will be compared.

To assess grid sensitivity, the root-mean-square deviation (RMSD) between experiment and simulation was computed for all six sensor locations according to

$$\text{RMSD} = \sqrt{\frac{\sum_{t=1}^{n}(T_{\text{exp},t} - T_{\text{sim},t})^2}{n}}, \quad \text{with } n = 60, \tag{13}$$

for $t \in \{1, 2, 3, \ldots, 60\,\text{min}\}$. Figure 7 confirms the generally good coincidence between simulations and measurements that was already evident from Figures 5 and 6. While the mean RMSD in the large tank decreased with a finer grid resolution, it remained almost constant in the small tank, which, on the other hand, showed a decrease in its standard deviation as the grid was refined. This indicates a systematic difference between simulations and experiments related with fluctuations of the experimental sensor data. For example, in the temperature measurements of T-F and T-N in Figure 6, a fluctuation between 18.2 °C and 18.3 °C was present during the entire cooling cycle. Also, it must be noted that the higher RMSD accuracy for the large tank scenario is clearly related to the fact that significant cooling effects showed up at only two of the six sensors compared to four of six positions in the small tank experiment. As explained, this is a consequence of the differences in the relative positions between cooling jacket and sensors. Again, this underlines the importance to consider the individual conditions in a winery when deciding on temperature sensor locations.

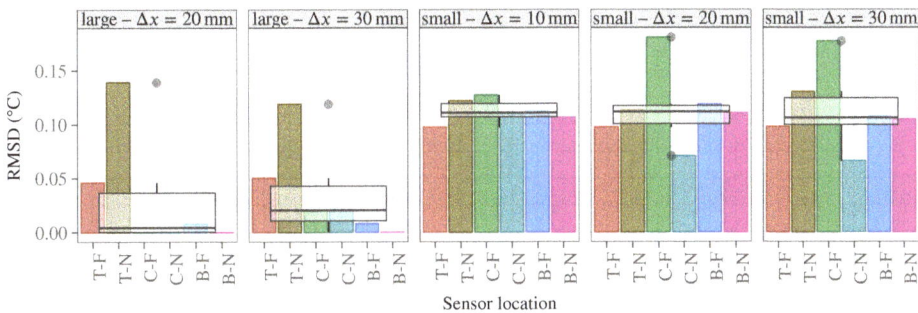

Figure 7. Root-mean-square deviation (RMSD) of liquid temperature between measurement and simulation for each sensor location and all case studies (large/small tank on different grid resolutions (Δx)). Box-plots represent across sensor location RMSD data for each study.

3.2. Sensor Location Scenarios

Finally, we analyze the software's suggestions on suitable sensor locations for the two case studies. In the large tank scenario—with the cooling jacket located deeply below the water surface—the software identifies a region of suitable sensor locations that lies slightly below the upper edge of the cooling jacket (Figure 8), based onsensor parameters $\Delta T_{sens} = 0.1\,°\text{C}$ and $q = 75\%$. On the finer grid, the region of suitable sensor locations is set a little lower compared to the coarser grid. Also, the finer grid suggests that a location in the central region of the tank might be less suitable compared to a location closer, but not too close, to the tank wall. From the coarser grid, on the other hand, no recommendations can be derived regarding the depth of suitable sensor locations, apart from the fact that the first few centimeters away from the cooling jacket are excluded. This corresponds to common practice where a minimum distance from the cooling jacket is always kept to avoid undesired measurement biases. Generally, it must be noted that in the large tank scenario the location of the cooling jacket is far from ideal w.r.t. temperature homogeneity. As experiments and the simulation in the previous section have shown, this configuration led to a distinct temperature stratification inside the tank which means that the use of mechanical mixers would be advisable.

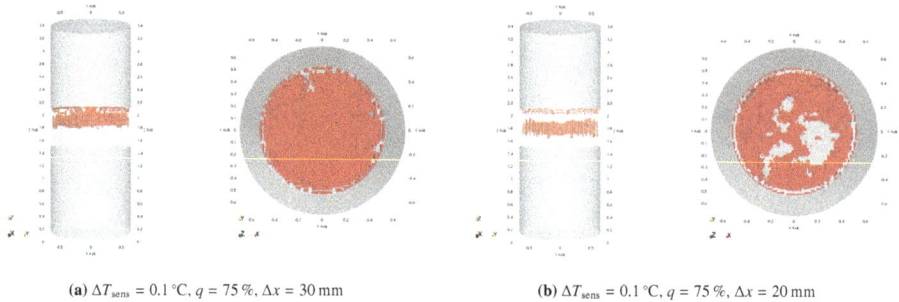

(a) $\Delta T_{sens} = 0.1\,°C$, $q = 75\,\%$, $\Delta x = 30\,mm$ \qquad (b) $\Delta T_{sens} = 0.1\,°C$, $q = 75\,\%$, $\Delta x = 20\,mm$

Figure 8. Post-processing output showing suitable sensor locations (red) for the large tank case study assessed for different Δx ((**a**): 30 mm; (**b**): 20 mm) and constant threshold values ($\Delta T_{sens} = 0.1\,°C$ and $q = 75\,\%$).

Less stratification was found in the small tank case study where the cooling jacket was located just slightly below the liquid surface. In this scenario, the differences between the regions of suitable sensor locations suggested by the software for each of the three grid resolutions were more pronounced, based on threshold parameters $\Delta T_{sens} = 0.1\,°C$ and $q = 75\,\%$ (Figure 9a–c). The results for the coarsest grid (Figure 9a, $\Delta x = 30$ mm) only excluded small portions at the bottom and top ot the tank from the suggested region of suitable sensor locations. The finer grids ($\Delta x = 10$ mm and 20 mm) identified the region of suitable sensor locations with more precision in a smaller volume, corresponding to about 1/3 to 2/3 of the liquids fill level. The larger size of the region of suitable sensor locations in comparison to the large tank scenario can be explained by the more homogeneous temperature distribution in the small tank. As discussed before, the grid refinements led to a more distinct temperature gradient between the two lower sensor levels which shrinks the size of the region of suitable sensor locations on the two finer grids relative to the coarsest grid.

By adjusting the threshold for the percentage of valid measurements to $q = 95.83\,\%$ (115/120 measurements) in the coarsest grid case, it was possible to shrink the region to a similar volume that was found in the simulations with the finer grid. This adjustment can be done in the post-processing step in ParaView, and hence the end-user can analyze effects of more strict thresholds after the simulation step to obtain smaller and more informative regions of suitable sensor locations. While results similar to the more computationally expensive fine grid simulations could be obtained on the coarse grid in this particular case, the end-user should be aware of the danger that the loss in accuracy on coarser grids may lead to unphysical results in other cases. The case studies suggest grid resolutions in a range between 10 mm to 30 mm for small to medium sized wine tanks as a reasonable compromise between computational costs and accuracy.

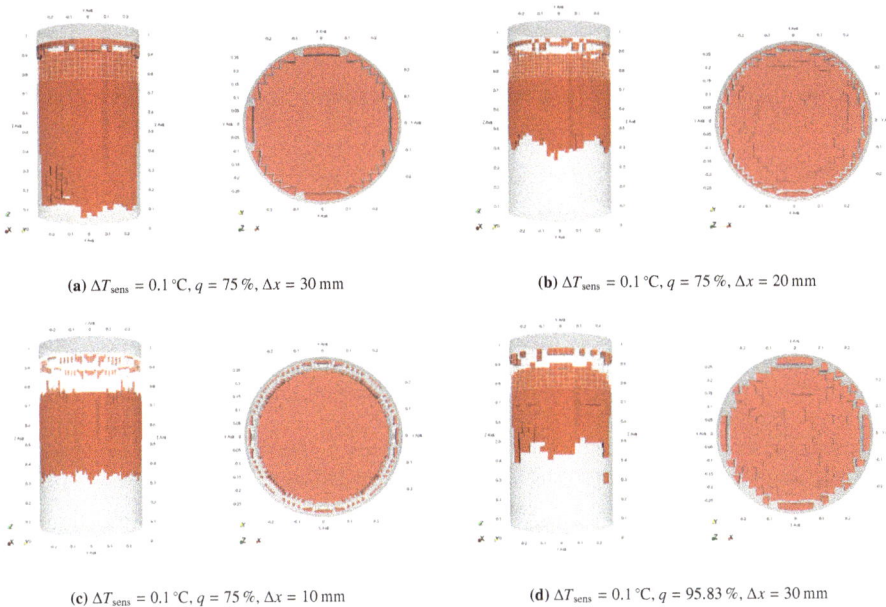

(a) $\Delta T_{sens} = 0.1\,°C$, $q = 75\,\%$, $\Delta x = 30\,mm$ (b) $\Delta T_{sens} = 0.1\,°C$, $q = 75\,\%$, $\Delta x = 20\,mm$

(c) $\Delta T_{sens} = 0.1\,°C$, $q = 75\,\%$, $\Delta x = 10\,mm$ (d) $\Delta T_{sens} = 0.1\,°C$, $q = 95.83\,\%$, $\Delta x = 30\,mm$

Figure 9. Post-processing output showing suitable sensor locations (red) for the small tank case study assessed for different Δx ((**a**): 30 mm; (**b**): 20 mm; (**c**): 10 mm) and constant threshold values ((**a–c**): $\Delta T_{sens} = 0.1\,°C$ and $q = 75\%$). (**d**) Effect of changing the percentage of valid measurements q to 95.83% for $\Delta x = 30\,mm$.

3.3. Computational Costs

Table 3 shows the computational costs for the case study simulations referring to an affordable standard workstation. As can be seen, computational times range between minutes to less than a day (1207 min), which means that it is feasible to perform the simulations with the software on standard end-user equipment, especially when choosing coarser grid resolutions. As it was shown, coarser grid simulations may already lead to a sufficiently good agreement with experimental data, while undesired effects of numerical diffusion can be reduced by lowering the threshold values. To avoid a too high loss in accuracy on coarser grids, a grid sensitivity study is recommended for tank sizes beyond the scope of this study, which can be easily realized in the end-user software by performing iterated simulations with several Δx values.

Table 3. Computational costs depending on grid size in the small and large tank scenarios.

	Δx	Grid Cells	Cores [a]	Simulation Time (min)	
				Real Time	Single Core
small	30 mm	12,805	3	3.73	11.19
	20 mm	40,154	18	8.92	160.56
	10 mm	310,252	18	157.25	2830.5
large	30 mm	214,036	3	729.47	2188.4
	20 mm	714,920	15	1207.35	21,735.54

[a] System: 2× Intel®Xeon®CPU E5-2660 v3 @ 2.60GHz–64GB RAM.

3.4. Summary of Results

A summary of our main results is given in Table 4.

Table 4. Summary of results.

Case studies	- Successful validation of CFD simulations against experimental temperature progression in 300 L and 5420 L jacketed wine tanks. - Pre-installed sensor locations not well suited to monitor bulk mean temperature.
Sensor locations	- Efficient sensor location depends on fill level and cooling jacket position. - More variability in height than in depth (distance to wall)
Computational costs	- Computational times of less than one day already lead to sufficient results.

4. Conclusions

An end-user software was developed that identifies optimized sensor locations for liquid bulk temperature measurement. Based on a few input parameters, different scenarios can be analyzed and optimized, or the suitability of pre-installed sensors can be evaluated. As it was shown, ideal temperature sensor locations in wine tanks depend strongly on the fill level relative to the location of the cooling device. Since the optimized sensor positions in our case studies showed a higher variability in height compared to depth (i.e. wall distance), an enhanced mobility of the sensors in height direction would be a useful enhancement for the wine industry. In this way, effective adaptive cooling strategies could be implemented based on the software that would be sensitive to variable conditions between the years, such as yield differences. The software could be used to compute optimal sensor height from live data, with subsequent adjustments of the physical sensors that would improve process control and efficiency. Based on a parameter study across a broad range of possible scenarios, the results could be formulated in terms of more general, phenomenological mathematical approaches such as Pareto models that would allow for a faster prediction of unevaluated scenarios [47,48]. This data base could also be used for a faster identification of a range of cooling scenarios that are appropriate for pre-installed, non-movable sensors.

Author Contributions: Conceptualization, D.S., M.F. and K.V.; Data curation, D.S.; Formal analysis, D.S. and K.V.; Funding acquisition, K.V.; Investigation, D.S. and M.F.; Methodology, D.S., M.F. and K.V.; Project administration, K.V.; Resources, D.S. and M.F.; Software, D.S.; Supervision, K.V.; Validation, D.S., M.F. and K.V.; Visualization, D.S.; Writing—original draft, D.S. and K.V.; Writing—review & editing, D.S., M.F. and K.V.

Funding: This research was funded by the German Federal Ministry of Education and Research (BMBF, Germany)—grant number 05M13RNA—as a part of the joint project "Robust energy optimization of fermentation processes for the production of biogas and wine (ROENOBIO)".

Conflicts of Interest: The authors declare no conflict of interest.

Abbreviations

The following abbreviations are used in this manuscript:

c_p	specific heat capacity ($\mathrm{J\,kg^{-1}\,K^{-1}}$)
d_T	tank diameter (m)
f_{Sed}	sediment fraction ($\mathrm{m^3\,m^{-3}}$)
g	gravitational acceleration ($\mathrm{m\,s^{-2}}$)
g_b	Boussinesq gravity ($\mathrm{m\,s^{-2}}$)
h_J	jacket height (m)
$h_{0,J}$	cooling jacket distance from tank height (m)
h_T	tank height (m)
p	pressure (Pa)
Pr	Prandtl number (-)
Pr_t	turbulent Prandtl number (-)

$\dot{Q}_a$	Rate of heat flow from ambient air to the liquid (W)
$\dot{Q}_c$	cooling power (W)
$\dot{Q}$	rate of heat flow (W)
q	percentage of valid measurements (%)
T	temperature (K)
T_x	cell center temperature (K)
$\dot{T}_c$	cooling rate (°C h^{-1})
ΔT_{sens}	accepted temperature tolerance (mm)
T_f	boundary face temperature (K)
$T_{0,1}$	initial fluid temperature (°C)
∇T_{lref}	local reference temperature gradient (K)
T_{Ref}	reference temperature (K)
Δt_{sens}	measurement interval of the temperature sensor (s)
t_{sim}	simulation duration (s)
u	velocity (m s^{-1})
V	liquid volume (m^3)
Δx	grid cell size (mm)
Greek symbols	
β	thermal expansion coefficient (K^{-1})
Δ_{fx}	face-to-cell distance (m)
κ	thermal conductivity (W m^{-1} K^{-1})
ν	kinematic viscosity (m^2 s^{-1})
ν_t	turbulent kinematic viscosity (m^2 s^{-1})
φ	cone angle (°)
ρ	density (kg m^{-3})
Subscripts	
c	linear cooling
f	linear face
L	linear large tank
S	linear small tank
x	linear cell
Acronyms	
CFD	computational fluid dynamics

References

1. Boulton, R.; Singleton, V.; Bisson, L.; Kunkee, R. Heating and Cooling Applications. In *Principles and Practices of Winemaking*; Springer: New York, NY, USA, 1999; pp. 492–520. [CrossRef]

2. Molina, A.; Swiegers, J.; Varela, C.; Pretorius, I.; Agosin, E. Influence of wine fermentation temperature on the synthesis of yeast-derived volatile aroma compounds. *Appl. Microbiol. Biotechnol.* **2007**, *77*, 675–687, doi:10.1007/s00253-007-1194-3. [CrossRef] [PubMed]

3. Masneuf-Pomarède, I.; Mansour, C.; Murat, M.L.; Tominaga, T.; Dubourdieu, D. Influence of fermentation temperature on volatile thiols concentrations in Sauvignon blanc wines. *Int. J. Food Microbiol.* **2006**, *108*, 385–390, doi:10.1016/j.ijfoodmicro.2006.01.001. [CrossRef] [PubMed]

4. Beltran, G.; Novo, M.; Guillamón, J.M.; Mas, A.; Rozès, N. Effect of fermentation temperature and culture media on the yeast lipid composition and wine volatile compounds. *Int. J. Food Microbiol.* **2008**, *121*, 169–177, doi:10.1016/j.ijfoodmicro.2007.11.030. [CrossRef] [PubMed]

5. Torija, M.; Rozès, N.; Poblet, M.; Guillamón, J.M.; Mas, A. Effects of fermentation temperature on the strain population of Saccharomyces cerevisiae. *Int. J. Food Microbiol.* **2003**, *80*, 47–53, doi:10.1016/S0168-1605(02)00144-7. [CrossRef]

6. Bisson, L.F. Stuck and Sluggish Fermentations. *Am. J. Enol. Vitic.* **1999**, *50*, 107–119.

7. Benitez, J.G.; Macias, V.P.; Gorostiaga, P.S.; Lopez, R.V.; Rodriguez, L.P. Comparison of electrodialysis and cold treatment on an industrial scale for tartrate stabilization of sherry wines. *J. Food Eng.* **2003**, *58*, 373–378, doi:10.1016/S0260-8774(02)00421-1. [CrossRef]

8. Jourdes, M.; Michel, J.; Saucier, C.; Quideau, S.; Teissedre, P.L. Identification, amounts, and kinetics of extraction of C-glucosidic ellagitannins during wine aging in oak barrels or in stainless steel tanks with oak chips. *Anal. Bioanal. Chem.* **2011**, *401*, 1531, doi:10.1007/s00216-011-4949-8. [CrossRef] [PubMed]

9. Forsyth, K.; Roget, W.; O'Brien, V. *Improving Winery Refrigeration Efficiency, Final Report, Project Number AWR 0902*; The Australian Wine Research Institute: Adelaide, Australia, 2012.

10. Soccol, C.R.; Pandey, A.; Larroche, C. *Fermentation Processes Engineering in the Food Industry*; CRC Press: Boca Raton, FL, USA, 2013.

11. Schmidt, D.; Velten, K. Numerical simulation of bubble flow homogenization in industrial scale wine fermentations. *Food Bioprod. Process.* **2016**, *100*, 102–117, doi:10.1016/j.fbp.2016.06.008. [CrossRef]

12. Zenteno, M.I.; Pérez-Correa, J.R.; Gelmi, C.A.; Agosin, E. Modeling temperature gradients in wine fermentation tanks. *J. Food Eng.* **2010**, *99*, 40–48, doi:10.1016/j.jfoodeng.2010.01.033. [CrossRef]

13. Vlassides, S.; Block, D.E. Evaluation of cell concentration profiles and mixing in unagitated wine fermentors. *Am. J. Enol. Vitic.* **2000**, *51*, 73–80.

14. Han, Y.; Wang, R.; Dai, Y. Thermal stratification within the water tank. *Renew. Sustain. Energy Rev.* **2009**, *13*, 1014–1026. [CrossRef]

15. Takamoto, Y.; Saito, Y. Thermal Convection in Cylindro-Conical Tanks During the Early Cooling Process. *J. Inst. Brew.* **2003**, *109*, 80–83, doi:10.1002/j.2050-0416.2003.tb00596.x. [CrossRef]

16. Meironke, H.; Kasch, D.; Sieg, R. Determining the thermal flow structure inside fermenters with different shapes using Ultrasonic Doppler Velocimetry. In Proceedings of the 10th International Symposium on Ultrasonic Doppler Methods for Fluid Mechanics and Fluid Engineering, Tokyo, Japan, 28–30 September 2016; pp. 125–128.

17. Schandelmaier, B. Gärsteuerung 2013—Ob groß oder klein—Pillow plates müssen sein. *Das Dtsch. Weinmagazin* **2013**, *20*, 30–36.

18. Morakul, S.; Mouret, J.R.; Nicolle, P.; Trelea, I.C.; Sablayrolles, J.M.; Athes, V. Modelling of the gas–liquid partitioning of aroma compounds during wine alcoholic fermentation and prediction of aroma losses. *Process Biochem.* **2011**, *46*, 1125–1131. doi:10.1016/j.procbio.2011.01.034. [CrossRef]

19. Galitzky, C.; Worrell, E.; Healy, P.; Zechiel, S. Benchmarking and self-assessment in the wine industry. In Proceedings of the 2005 ACEEE Summer Study on Energy Efficiency in Industry, West Point, NY, USA, 19–22 July 2005.

20. Perez-Coello, M.; Gonzalez-Vinas, M.; Garcia-Romero, E.; Diaz-Maroto, M.; Cabezudo, M. Influence of storage temperature on the volatile compounds of young white wines. *Food Control* **2003**, *14*, 301–306, doi:10.1016/S0956-7135(02)00094-4. [CrossRef]

21. Joyeux, A.; Lafon-Lafourcade, S.; Ribéreau-Gayon, P. Evolution of acetic acid bacteria during fermentation and storage of wine. *Appl. Environ. Microbiol.* **1984**, *48*, 153–156. [PubMed]

22. Lafon-Lafourcade, S.; Carre, E.; Ribéreau-Gayon, P. Occurrence of lactic acid bacteria during the different stages of vinification and conservation of wines. *Appl. Environ. Microbiol.* **1983**, *46*, 874–880. [PubMed]

23. Boquete, L.; Cambralla, R.; Rodríguez-Ascariz, J.; Miguel-Jiménez, J.; Cantos-Frontela, J.; Dongil, J. Portable system for temperature monitoring in all phases of wine production. *ISA Trans.* **2010**, *49*, 270–276, doi:10.1016/j.isatra.2010.03.001. [CrossRef] [PubMed]

24. Zhang, W.; Skouroumounis, G.K.; Monro, T.M.; Taylor, D. Distributed Wireless Monitoring System for Ullage and Temperature in Wine Barrels. *Sensors* **2015**, *15*, 19495–19506, doi:10.3390/s150819495. [CrossRef] [PubMed]

25. Cañete, E.; Chen, J.; Martín, C.; Rubio, B. Smart Winery: A Real-Time Monitoring System for Structural Health and Ullage in Fino Style Wine Casks. *Sensors* **2018**, *18*, 803, doi:10.3390/s18030803. [CrossRef] [PubMed]

26. Ranasinghe, D.C.; Falkner, N.J.; Chao, P.; Hao, W. Wireless sensing platform for remote monitoring and control of wine fermentation. In Proceedings of the 2013 IEEE Eighth International Conference on Intelligent Sensors, Sensor Networks and Information Processing, Melbourne, VIC, Australia, 2–5 April 2013; pp. 503–508.

27. Sainz, B.; Antolín, J.; López-Coronado, M.; Castro, C.D. A Novel Low-Cost Sensor Prototype for Monitoring Temperature during Wine Fermentation in Tanks. *Sensors* **2013**, *13*, 2848–2861, doi:10.3390/s130302848. [CrossRef] [PubMed]

28. Di Gennaro, S.F.; Matese, A.; Mancin, M.; Primicerio, J.; Palliotti, A. An Open-Source and Low-Cost Monitoring System for Precision Enology. *Sensors* **2014**, *14*, 23388–23397, doi:10.3390/s141223388. [CrossRef] [PubMed]

29. Ferziger, J.H.; Peric, M. *Computational Methods for Fluid Dynamics*; Springer Science & Business Media: New York, NY, USA, 2012.

30. Weller, H.G.; Tabor, G.; Jasak, H.; Fureby, C. A tensorial approach to computational continuum mechanics using object-oriented techniques. *Comput. Phys.* **1998**, *12*, 620–631, doi:10.1063/1.168744. [CrossRef]

31. Rodríguez, I.; Castro, J.; Pérez-Segarra, C.; Oliva, A. Unsteady numerical simulation of the cooling process of vertical storage tanks under laminar natural convection. *Int. J. Therm. Sci.* **2009**, *48*, 708–721, doi:10.1016/j.ijthermalsci.2008.06.002. [CrossRef]

32. Papanicolaou, E.; Belessiotis, V. Transient natural convection in a cylindrical enclosure at high Rayleigh numbers. *Int. J. Heat Mass Trans.* **2002**, *45*, 1425–1444, doi:10.1016/S0017-9310(01)00258-7. [CrossRef]

33. R Core Team. *R: A Language and Environment for Statistical Computing*; R Foundation for Statistical Computing: Vienna, Austria, 2017.

34. Chang, W.; Cheng, J.; Allaire, J.; Xie, Y.; McPherson, J. Shiny: Web Application Framework for R; R Package Version 1.0.3; 2017. Available online: https://cran.r-project.org/package=shiny (accessed on 24 May 2018).

35. CEA/DEN, EDF R&D, O.C. SALOME Geometry User's Guide; CEA/DEN, EDF R&D, OPEN CASCADE; 2016. Available online: http://docs.salome-platform.org/7/gui/GEOM/ (accessed on 24 May 2018).

36. Ahrens, J.; Geveci, B.; Law, C. *ParaView: An End-User Tool for Large Data Visualization*; Visualization Handbook; Elsevier: Amsterdam, The Netherlands, 2005.

37. Günther, M.; Velten, K. *Mathematische Modellbildung und Simulation*; Wiley-VCH: Berlin, Germeny, 2014.

38. Velten, K.; Müller, J.; Schmidt, D. New methods to optimize wine production at all stages from vineyard to bottle. *BIO Web Conf.* **2015**, *5*, 02013, doi:10.1051/bioconf/20150502013. [CrossRef]

39. Wickham, H. *Ggplot2: Elegant Graphics for Data Analysis*; Springer: New York, NY, USA, 2009.

40. Urbanek, S.; Horner, J. Cairo: R Graphics Device Using Cairo Graphics Library for Creating High-Quality Bitmap (PNG, JPEG, TIFF), vector (PDF, SVG, PostScript) and Display (X11 and Win32) Output; R Package Version 1.5-9; 2015. Available online: https://cran.r-project.org/package=Cairo (accessed on 24 May 2018).

41. Bengtsson, H. Future: Unified Parallel and Distributed Processing in R for Everyone; R Package Version 1.5.0; 2017. Available online: https://cran.r-project.org/package=future (accessed on 24 May 2018).

42. Greenshields, C. *OpenFOAM User Guide*, v5 ed.; CFD Direct Ltd.: Reading, UK, 2017. Available online: https://cfd.direct/openfoam/user-guide/ (accessed on 24 May 2018).

43. De Moura, M.D.; Júnior, A.C. Heat Transfer by Natural ConvectIon in 3D Enclosures. In Proceedings of the ENCIT 2012, Rio de Janeiro, Brazil, 18–22 November 2012.

44. Corzo, S.F.; Ramajo, D.E.; Nigro, N.M. CFD model of a moderator tank for a pressure vessel PHWR nuclear power plant. *Appl. Therm. Eng.* **2016**, *107*, 975–986. doi:10.1016/j.applthermaleng.2016.07.034. [CrossRef]

45. McQuillan, F.J.; Culham, J.R.; Yovanovich, M.M. *Properties of Some Gases and Liquids at One Atmosphere*; Technical Report; Microelectronics Heat Transfer Laboratory Report UW/MHTL 8407 G-02; Microelectronics Heat Transfer Laboratory, University of Waterloo: Waterloo, ON, Canada, 1984.

46. Wagner, W.; Pruß, A. The IAPWS Formulation 1995 for the Thermodynamic Properties of Ordinary Water Substance for General and Scientific Use. *J. Phys. Chem. Ref. Data* **2002**, *31*, 387–535, doi:10.1063/1.1461829. [CrossRef]

47. Velten, K. *Mathematical Modeling and Simulation: Introduction for Scientists and Engineers*; John Wiley & Sons: New York, NY, USA, 2009.

48. Safikhani, H.; Hajiloo, A.; Ranjbar, M. Modeling and multi-objective optimization of cyclone separators using CFD and genetic algorithms. *Comput. Chem. Eng.* **2011**, *35*, 1064–1071, doi:10.1016/j.compchemeng.2010.07.017. [CrossRef]

MDPI

St. Alban-Anlage 66

4052 Basel

Switzerland

Tel. +41 61 683 77 34

Fax +41 61 302 89 18

www.mdpi.com

Fermentation Editorial Office

E-mail: fermentation@mdpi.com

www.mdpi.com/journal/fermentation

www.ingramcontent.com/pod-product-compliance
Lightning Source LLC
Chambersburg PA
CBHW051859210326
41597CB00033B/5959